Designed for Play

New Historical Perspectives is an open access book series for early career scholars, commissioned, edited and published by the Royal Historical Society and the University of London Press in association with the Institute of Historical Research. Submissions are encouraged relating to all historical periods and subjects. Books in the series are overseen by an expert editorial board to ensure the highest standards of peer-reviewed scholarship, and extensive support and feedback for authors is provided.

The series is supported by the Economic History Society and the Past and Present Society.

Recently published

Gender, Emotions and Power, 1750–2020, edited by Hannah Parker and Josh Doble (November 2023)
Anti-Communism in Britain During the Early Cold War, by Matthew Gerth (April 2023)

Designed for Play

Children's Playgrounds and the Politics of Urban Space, 1840–2010

Jon Winder

Available to purchase in print or download for free at
https://uolpress.co.uk/

First published 2024 by
University of London Press
Senate House, Malet St, London WC1E 7HU

A CIP catalogue record for this book is available from The British Library.

ISBN 978-1-914477-48-5 (hardback)
ISBN 978-1-914477-49-2 (paperback)
ISBN 978-1-914477-51-5 (.epub)
ISBN 978-1-914477-50-8 (.pdf)
ISBN 978-1-914477-68-3 (.html)

DOI https://doi.org/10.14296/mgyc2910

Cover image: Children Playing at the Foundlings Site Playing Fields by Marshall, 1936 © Daily Herald Archive / Science Museum Group

Cover design for University of London Press by Nicky Borowiec.
Series design by Nicky Borowiec.
Book design by Nigel French.
Text set by Westchester Publishing Services UK in Meta Serif and Meta, designed by Erik Spiekermann.

For Alix, Edie and Jake

Contents

List of figures

Preface

... all historians are quirky individuals first, readers second, and writers third.[1]

Perhaps the most significant individual quirk that frames *Designed for Play* is fifteen years spent working for numerous local authorities on the restoration of historic parks and the creation of spaces for children to play. When I first started, it seemed obvious that playgrounds were for children. But over time I experienced an increasing uncertainty about who playgrounds were for and what purpose they served. In trying to work this out, I talked to lots of people – and they all gave me different answers. Children modelled, sketched and talked about expansive tree-houses, underground tunnels, mid-air swimming pools, huge slides, secluded hideaways, water jets and other spaces of imagination and excitement. Busy parents and carers often requested a sturdy boundary fence, to keep dogs out and their children in. Maintenance staff preferred robust metal structures (and definitely no sand or water). Local politicians tended to be most interested in the publicity photo at the end of the project, ready for the next election campaign. Some advocates saw playgrounds as spaces where children could exercise power and influence over their play experiences, whiles others argued that we should be focusing on the wider urban environment and the creation of child-friendly cities more generally.

In response, we installed sandpits, tree swings, water features and treehouses. But it soon became clear that playgrounds were not the simple spaces for children's play that I had initially imagined. If they were the right approach to meeting the needs of children, what social and environmental problems were they seeking to address? If they were not the right approach, then why did we spend so much time and money creating such spaces and attempting to care for them? Unfortunately, neither scholarly research nor popular accounts of the playground provided satisfactory answers to these questions. A brief foray into the writing of playground advocate Marjory Allen and urban commentator Jane Jacobs provided some critical and compelling leads, but also highlighted the need for a dedicated historical study.

A chance meeting with Karen Jones, in the picturesque grounds of Kearsney Abbey in Kent, reintroduced me to the possibilities of academic research and provided the spark that inspired this historical playground project. I am extremely grateful for Karen's intellectual encouragement (and practical support for numerous funding bids). Since then, a supportive community of friends and scholars have been tirelessly encouraging, most notably Joe Jones, but also Ben Highmore, Charlotte Sleigh, Juliette Pattinson, Clare Hickman and a pan-European network of playground historians. Generous financial support from the CHASE doctoral training partnership, a Royal Historical Society Early Career Fellowship grant and a Scouloudi Foundation publication grant (via the Institute of Historical Research) have been invaluable. I am also indebted to the supportive team behind the New Historical Perspectives series at the University of London Press, especially Elizabeth Hurren and Emma Gallon.

Most importantly, Edie and Jake's playful exploration of playgrounds, parks, beaches and forests has been fun and inspiring in equal measure. Our quirky adventures have sustained me throughout this project, thank you so much.

Notes

1. Susan A. Crane, 'Historical Subjectivity: A Review Essay', *The Journal of Modern History*, 78.2 (2006), 434–56 (p. 435).

Introduction

The children's playground is a ubiquitous feature of British towns and cities. Such spaces, with their swings and roundabouts, are often seen as the obvious place for children to play: safe, natural and out of the way. But these assumptions hide a previously overlooked history of children's place in public space, one shaped by an inequitable distribution of power and implicit assumptions about age, gender, class and the environment. Perhaps surprisingly, given their ubiquity and our near universal experience of using them, the provision of dedicated places for play has not been required by law or prescribed by central government. Instead, the erratic evolution of the playground in Britain has been shaped by competing responses to the social and environmental problems of the industrial city, across the nineteenth and twentieth centuries. In the past, public park advocates, housing reformers, ardent imperialists, committed anarchists, municipal authorities and philanthropic industrialists have all embraced the potential of the children's playground to deliver wider social, political and commercial ambitions. As a result, the playground has become firmly embedded in both imagined and material urban landscapes.

At the same time, we instinctively know that children play everywhere and with everything. Young children can often be found playing with cardboard boxes and kitchen utensils, alongside the toys and games specifically designed for play. The walk to school is frequently a meandering journey of imaginative adventure and splashing in puddles, while the beach and its messy combination of water, sand and relative freedom is a playful favourite. However, if these examples are typical, why do we provide specific public spaces where children are supposed to play and what purpose are such spaces meant to serve? Today, we might suggest that anxiety about stranger danger or the lethal threat posed by motor vehicles provide strong justification for creating and maintaining playgrounds. However, the earliest dedicated public spaces for play pre-date both the invention of the car and more recent anxiety about the threat to children from strangers, suggesting a far more complex story.

Designed for Play asks vital questions about the apparently commonsense association between children and the playground. It provides an essential point of reference for scholars, policymakers and campaigners seeking to understand and enhance children's place in the social and physical worlds. The book exposes the enduring tension between children's universal desire to play and adult attempts to influence and direct such playfulness. Using a wide range of previously unexamined archive materials, it offers a unique account of the children's gymnasium and giant stride, joy wheel and ocean wave, multimillion-pound philanthropic donations and the utopian visions of pioneering playground advocates. It makes a significant contribution to our understanding of the diverse historical and geographical themes that have shaped both public childhoods and public spaces, transcending conventional academic boundaries.

In doing so, it finds a convoluted history, one where the form and function of play spaces have long reflected adult anxiety about urban childhood, rather than necessarily the needs or preferences of children at play. For over 150 years, the children's playground has represented a space where changing conceptions of urban childhood, nature, health and commerce have all been played out. However, the politics and values that have informed playground creation have rarely been considered by academics, professionals or the wider public. This in turn has resulted in present-day uncertainty about the social and political purpose of designated spaces for children's playful recreation. To address this, *Designed for Play* adopts a long chronology to explore how anxieties about childhood and the urban environment have intermittently converged, with lasting consequences for both children and cities. For the first time, it uncovers the main actors involved and examines the assumptions, motivations and wider historical themes that have shaped play provision on the ground over the course of two centuries. In doing so, it positions children more centrally in our understanding of both the nineteenth-century parks movement and twentieth-century visions for the modern urban environment, and tracks the fluctuating significance of philanthropy, voluntary action, state intervention and commerce in shaping both public life and public space.

Playgrounds today

While a trip to the playground might be a fun and seemingly playful venture, behind the scenes present-day play spaces are an important site of social and political contest, generating considerable scholarly and public discourse about children's place in both social and physical worlds. From Auckland and Athens to Singapore and Seattle, scholars have debated the problems and possibilities of the playground. Researchers have variously asserted that such bounded spaces are a symbol of children's inequitable access to the city, a spatial predictor of adolescent drug use or sites for the powerful expression of playful child agency.[1] In the UK, £86m is spent each year maintaining more than 26,000 public playgrounds, but with little strategic direction about their purpose and form.[2] This uncertainty about the relationship between place and play has also regularly featured in creative expression, as artists have sought to make sense of the playground and its social function. In the last decade, exhibitions at the Museum of Modern Art in New York, the Royal Institute of British Architects in London, the Kunsthalle in Zurich and the Baltic Centre for Contemporary Art in Gateshead have all sought to problematise established ideas about the place of children's play in the city.[3]

Beyond academia and the arts, the playground has also been a focal point for popular anxiety about the apparent disconnect between twenty-first-century childhood and the 'natural' environment. From impassioned pleas by naturalists and calls for a rewilding of childhood, to sobering statistics about the limits of youthful interaction with the natural world, there is considerable concern about children's separation from 'nature' and a renewed emphasis on the need for more natural play spaces.[4] Although these widely held assumptions about childhood and nature are not unproblematic, they nonetheless exert a significant influence on present-day debate about play space design.[5] Alongside this anxiety about urban nature, there have also been increasing calls to rethink urban infrastructure from a child's point of view. From UNICEF's Child Friendly Cities Initiative to the work of pioneering municipal authorities, a child-centric and family-friendly approach to the planning of housing, transport and the wider public realm is increasingly positioned as an essential part of creating inclusive urban environments.[6] Meanwhile, aristocratic landowners have created substantial playgrounds, with high entrance fees, as part of wider efforts to generate income to help maintain their estates.[7] Together, this complex range of responses to the present-day playground point to the need for an empirically grounded historical study, one that will help to make sense of the shifting values and assumptions that have shaped the enduring provision and contested form of dedicated places for play.

Playing in the past

This interest in the social purpose of children's play is hardly a new phenomenon. Children have always played and as adults we have long sought to direct such playfulness, asserting that youthful recreation might perform a useful function, often prescribing where it should take place. Evidence that play has long been an important feature of childhood can be seen in museums around the world, from 4,000-year-old Egyptian and Indus Valley toys to ancient Greek ornaments depicting children playing games with friends.[8] The captivating early modern oil painting *Children's Games* (1560) by Pieter Bruegel depicts in encyclopaedic detail over 200 children playing in an imagined Dutch townscape.[9] Over eighty playful activities are shown taking place in outdoor spaces, encompassing the urban street and town square, rural fields and nearby stream, reflecting the contemporary attitudes of the Dutch mercantile classes towards play and education.[10]

Beyond the museum and gallery, historians have found considerable evidence of child-specific toys, games and equipment, including a medieval dictionary entry for a 'merrytotter', an undefined structure seemingly intended to encourage children's play outside, perhaps comparable to a swing or see-saw.[11] Although access to such structures would most likely have been constrained by children's age, gender and social status, the quality of spaces for youthful recreation has also long been important. According to Thomas Elyot, writing in 1531, the Romans set aside a large open space, the *Campus Martius*, next to the River Tiber so that children could exercise and play in the water.[12] For Elizabethan pedagogue Richard Mulcaster, firm ground, shelter from the 'byting winde' and fresh air that was free from a 'noisome stenche' were essential features of a ground for the physical education of children.[13] These associations between notions of childhood and environment would continue well into the nineteenth century. James Kay-Shuttleworth, the noted Victorian educationalist and Poor Law Commissioner, felt that an appropriately laid out playground provided a 'means of teaching the children to play without discord'.[14]

Charting a longer history of play spaces for children is complicated by the ambiguous epistemology of the word 'playground'. Today, the *Oxford English Dictionary* defines the term as either 'a piece of ground used for playing on' or in extended use 'any place of recreation'.[15] When imagining the playground today, we tend to think of it as a place for children to play, most likely equipped with swings, slides and other equipment. However, the term has not always been used to describe spaces specifically set aside for children. In 1768, the physician Francis de Valangin used the term 'play-ground' to describe a green open area for curative recreation beyond London's city wall.[16] In the 1830s, the residents of Hathern, Leicestershire, used the term to describe a place for playing sport, while for the noted mountaineer and author Leslie Stephen, the European Alps were the 'playground of Europe'.[17] In 1858, the successful Liverpool merchant Charles Melly used the term 'playground' to describe a space specifically for energetic exercise. He defined his free outdoor gymnasium as a playground for the healthful enjoyment of the city's working-class residents, but still not specifically for children.[18] In other contexts, the playground represented a space of education and rest for children but excluded the wider public. In the late eighteenth century, the travel writer Arthur Young described visiting an attractive school with a 'spacious playground walled in', while a boarding school for young gentleman in Ilford similarly included a large, enclosed playground and garden.[19] At the same time, the exclusion of the public from the school playground was sometimes problematic. In nineteenth-century Hackney, the enclosure of common land to create a playground for the Grocer's Company school resulted in a long running and high-profile dispute with the local community over Lammas rights.[20] As such, during the eighteenth and nineteenth centuries the term 'playground' was broadly conceived, variously used to represent spaces for education, exercise, sport and recreation. It also operated at a variety of scales and crossed the boundaries of public and private, childhood and adulthood.

In the face of such diverse meanings, the central focus here is on the public realm and the provision of dedicated public spaces for children's playful recreation, rather than school playgrounds or spaces primarily for adult use. In adopting such a stance, the book remains sensitive to the wider meaning that the term playground could represent, particularly its associations with education, health, exercise and adventure. A similarly flexible and sensitive approach is adopted in relation to the age-related boundaries of childhood. For much of the period in question, definitions shifted as legislation and social norms sought to shape the age at which sexual consent, education and work could take place. Playground advocates adopted similarly flexible definitions of childhood, rather than an absolute age range for the spaces they sought to create. Broadly speaking, advocates imagined that playgrounds would be used by children from toddlerhood, perhaps with older siblings, up to their teens and a similar approach is adopted here.

Playground histories

The longstanding interest in the spaces and social function of children's play might suggest a similarly enduring attention from the scholarly community. However, the playground has seldom been a feature of academic research, despite an increasing interest in the history of spaces where playgrounds are often found, such as public parks and housing estates, and studies of the enduring connection between landscapes and health. Despite undoubtedly being shaped by notions of education, welfare, health and leisure, the public playground has rarely been a feature of the historiography that covers these fields. Furthermore, attempts to construct playground narratives from beyond the discipline of history have either failed to justify their claims with historical evidence, and as a result have tended to overly romanticise the past, or have in turn relied on such unsubstantiated accounts to make their case. Moreover, there has been a tendency in popular accounts to present the history of the playground in the USA as a universal story that can be applied elsewhere. While there was undoubtedly an international exchange of ideas (something explored later in the book), the creation of dedicated spaces for children's public recreation was also reliant on local attitudes to childhood, urbanisation and a host of other factors, something clearly evidenced in historical scholarship on playground spaces in Budapest, Dublin, Helsinki and Toronto.[21]

That does not diminish the considerable historiography that deals with the story of the playground in the USA, with historical accounts published as early as 1922.[22] Subsequent studies have focused on the late nineteenth and early twentieth centuries, on the campaigning of the Playground Association of America (1906) and practical action in cities such as San Francisco and Cambridge, Massachusetts. Within these spatial and chronological boundaries, scholars have positioned the playground as a space where notions of gender, race and citizenship were negotiated and urban land politics were played out.[23] A notable development in the field in the USA has been research into the microhistories of individual playground sites, including the Hull House playground in Chicago, to better understand children's experiences of using them and contesting their boundaries.[24] That this research is possible points to a key difference between the development of playgrounds in Britain and the USA. In the latter, playgrounds were often highly organised spaces that involved significant adult organisation and coordination, with administrative records at individual playground sites preserved and archived.

In contrast, the promotion and management of playgrounds in Britain has involved a wide range of philanthropic, voluntary and governmental organisations whose remit often extended well beyond spaces for play. As a result, the playground archive is significantly more fragmented and dispersed. The name of the Metropolitan Public Gardens, Boulevard and Playground Association (1882) hints at its diverse campaigning interests, while its archive is spread around the world, with no central record of remaining materials. Within local government, playgrounds have variously been the responsibility of park superintendents, engineers, architects, surveyors and housing officers, and the quality and extent of record keeping has varied significantly across more than four hundred local authorities. The retention of records at Wicksteed Park has occurred somewhat by accident, and while a project is in the process of reviewing materials, there is still much to do to organise and catalogue records. As we might expect, the playground appears in more published material in the twentieth century, but still in no central repository. As such, the research that underpins *Designed for Play* has drawn on records and materials at the National Archives, London Metropolitan Archives, British Library, Wicksteed Park, the Royal Horticultural Society's Lindley Library, the Landscape Archives at the Museum of English Rural Life, the Royal Institute of British Architects, various university libraries and local authority record centres.

Partly as a result, the historiography on play space developments in Britain has tended to focus on radical playground experiments in the mid-twentieth century, often using published accounts of the activists involved. Valerie Wright's study of children's play on Glasgow council estates in the 1960s is a useful addition to historical scholarship which has otherwise mainly concentrated on the iconic mid-century adventure playground in bomb-damaged cities such as London, Liverpool and Bristol.[25] Historians, including Krista Cowman and Roy Kozlovsky, have explored the assumptions, values and practical action that shaped the adventure playground movement and its role in postwar reconstruction.[26] Cultural historians, including Ben Highmore and Lucie Glasheen, have also analysed postwar representations of children at play in the city and their wider cultural significance.[27] Drawing on both similar sources and new ones, later chapters in the book will provide a re-reading of this twentieth-century material to assess the significance of ideas about nature and education, as well as adventure playground thinking, on the wider provision of public places for play. In doing so, the book responds to Katy Layton-Jones's call for a study of landscapes designed for children, but also extends the chronology back into the nineteenth century to make sense of the wider social and environmental processes that shaped playground provision more broadly.[28]

Childhood and the urban environment

In narrating a more expansive historical account of the playground, *Designed for Play* builds on a broad consensus among historians that conceptions of childhood underwent significant change during the nineteenth century. In that period, childhood was increasingly imagined and constructed as a distinct phase of life, one that contrasted sharply with adulthood. The gradual expansion of compulsory education, alongside a broader focus on understanding children's minds and bodies, formed part of wider philanthropic and state-sponsored welfare directed at children. The principle of providing dedicated public places for play was undoubtedly influenced by broader attempts to extend childhood education in the nineteenth century, notably the increasing importance of outdoor games in schools, and efforts to reform educational provision in the twentieth century, inspired by radical pedagogy and the open-air school movement. However, the extension of these values into the public realm and their relationship with ideas about nature and the city has rarely been considered. Inspired by modern historians of childhood, *Designed for Play* highlights the socially, spatially and historically constructed nature of childhood, the extension of these values into the public realm and their intersection with ideas about nature and the city.[29] Drawing on historical photographs, newspaper articles and records which mention youthful activities, the book also shows how children adapted and contested adult expectations of the playground. In doing so, it seeks to balance on the one hand providing an original account of the adult anxieties, assumptions and practical action that led to the creation of playgrounds, while on the other hand being sensitive to examples of negotiation and contestation by children.

Just as attitudes to childhood and education were shifting in the nineteenth century, so too were concerns about the industrialised city and the impact of the urban environment on the physical and moral wellbeing of its inhabitants. Public parks, institutional gardens and other urban greenery were positioned as spaces of individual health and an environmental tonic for the ills of the city. However, there is often only a passing reference to amenities provided for children in historical accounts of this green infrastructure movement. Where children are mentioned in park histories, for instance, coverage tends to be cursory and often assumes that the place of play within the park is an obvious and natural one.

As will become clear in later chapters, this has not always been the case. Inspired by the work of social and environmental historians who have critically engaged with the eco-cultural values that have shaped public parks on both sides of the Atlantic, *Designed for Play* unpicks these assumptions about children's place in the park landscape.[30] It also considers the place of children in environmental histories, addressing their position as 'academic orphans' in the field.[31]

Beyond the park boundary, the fate of the playground would be closely tied to wider responses to the problems of the industrial city, both pragmatic and utopian. While the focus here is on the evolution of the children's playground, *Designed for Play* speaks more broadly to the increasing interest among philanthropic, voluntary and state actors in adapting the urban environment to achieve social, political and environmental ambitions during the nineteenth and twentieth centuries. In exploring these processes, the book contributes to recent interdisciplinary interest in narrating urban social histories and exploring the complex geographies and histories associated with ostensibly natural spaces and apparently biological assumptions about urban inhabitants.[32] Uniquely, in seeking to identify the origins of the playground in the mid-nineteenth century and its fortunes through to the early twenty first century, it extends the chronological scope of historiography that considers the spatial expression of urban modernisms, charting their fortunes over the course of 150 years.[33]

Designed for Play also contributes to revisionist accounts of welfare in Britain, particularly those that situate children and public space within such narratives. It points to the evolutionary and increasingly holistic nature of welfare provision, rather than sudden state involvement from 1945. Notably, this study expands on the chronological coverage provided by Mathew Thomson in *Lost Freedom* and shows that the provision of dedicated public spaces for children began much earlier than the 1940s.[34] In doing so, it provides important historical context for Thomson's work, suggesting that the perceived need to protect children from both the street and inappropriate adult behaviour, and the associated provision of special places for play, had much earlier roots than has generally been acknowledged. While *Lost Freedom* sets the stage for this study, here the focus shifts from an emphasis on the child to an interest in the social and spatial consequences of ideas about urban childhood, in particular for the provision, design and management of public spaces set aside for children. In adopting such an approach, *Designed for Play* contributes to our understanding of the spatial consequences of modern welfare, uncovering the ambition, design and policy that sought to shape urban landscapes and children's lived experience. It charts how administrators, professionals, academics and philanthropists sought to adapt the public realm in line with these ideas. But, as historian James Greenhalgh notes, such attempts to create more rational urban spaces and regulated behaviours were always negotiated by users and, as we shall see, children were particularly effective at subverting adult expectations about where and how they should play.[35]

Overview

For the first time, *Designed for Play* traces the untold story of the children's playground, from the mid-nineteenth century to the early twenty-first century. Taking in public parks, modern housing estates and other urban spaces, the book charts the playground's journey from marginal obscurity to popular ubiquity and more recent challenges to its status as a site of health, nature and safety. Organised around a chronological structure, it examines the wider social, political and environmental assumptions that shaped the creation of dedicated places for play, drawing on the archival materials of reformers, parks superintendents, equipment manufacturers and architects, in Britain and beyond.

Chapter 1 focuses on the nineteenth-century experience and the ameliorative potential of green space and exercise for unhealthy urban childhoods. It shows how dedicated spaces for children were seen by some as a way to mitigate the social and environmental consequences of the industrial city, but that efforts to create such spaces would be largely unsuccessful until conceptions of childhood also included time for leisure later in the century. Chapter 2 goes on to analyse how the principle of the children's playground became more firmly embedded in imagined and material urban landscapes as philanthropic, voluntary and state actors negotiated interventions into public space. The chapter examines the competing visions for the playground that were in circulation in the early twentieth century and the influence of commercial equipment manufacturers, particularly Charles Wicksteed & Co., in defining what would become the orthodox playground of swings, slides and roundabouts.

After this specific case study, Chapter 3 considers how this ideal type spread to cities across Britain in the interwar period, particularly as one solution to the dangers facing children when playing in the street. The increasing number of playgrounds and standardised design reflected municipal confidence in adapting the urban environment and the ongoing role of voluntary organisations in advocating for play. The chapter shows how swings in particular came to dominate playground spaces and charts the progress of wider debates about the role of adults in guiding children's play. Chapter 4 investigates how this orthodoxy, centred on manufactured playground equipment, was initially consolidated and then challenged in the mid-twentieth century, as campaigners inspired by international exemplars and the adventure playground movement sought to promote greater freedom and creativity in children's play. The chapter explores the work of pioneering play space advocates, including Marjory Allen's intervention in play space debate and the significance of her environmental biography, as well as the work of designer Mary Mitchell. Chapter 5 focuses on the later twentieth and early twenty-first century, plotting a battle for ideas in playground discourse and highlighting a number of challenges to perceptions of the playground as a safe and healthy space for children. It plots the fluctuating interest of central government in play space provision and the ongoing influence of equipment manufacturers in shaping both popular and professional notions of the ideal playground. It considers the contested place of the playground in local politics, sociological research and anarchic thought, before charting popular and political anxiety about playground safety. In adopting a long chronology and broad scope, *Designed for Play* makes an important contribution at the intersection of urban and environmental histories and the geographies and histories of childhood.

Notes

1. M. Kotlaja, E. Wright and A. Fagan, 'Neighborhood Parks and Playgrounds: Risky or Protective Contexts for Youth Substance Use?', *Journal of Drug Issues*, 48.4 (2018), 657–75; P. Carroll and others, 'A Prefigurative Politics of Play in Public Places: Children Claim Their Democratic Right to the City through Play', *Space and Culture*, 22.3 (2019), 294–307; A. Pitsikali and R. Parnell, 'Fences of Childhood: Challenging the Meaning of Playground Boundaries in Design', *Frontiers of Architectural Research*, 9.3 (2020), 656–69; R. Sini, 'The Social, Cultural, and Political Value of Play: Singapore's Postcolonial Playground System', *Journal of Urban History*, 48.3 (2022), 578–607.

2. Jon Winder, 'Children's Playgrounds: "Inadequacies and Mediocrities Inherited from the Past"?', *Children's Geographies*, 2023, 1–6 https://doi.org/10.1080/14733285.2023.2197577.

3. Juliet Kinchin and Aidan O'Connor, *Century of the Child: Growing by Design 1900–2000* (New York: The Museum of Modern Art, 2012); Simon Terrill and Assemble, *The Brutalist Playground*, 2015, RIBA; Burkhalter, Gabriela, ed., *The Playground Project* (Zurich: JRP|Ringier, 2016); Albert Potrony, *Equal Play*, 2021, BALTIC.

4. Natural England, *The Children's People and Nature Survey for England* (London: Office for National Statistics, 2022).

5. Elizabeth Dickinson, 'The Misdiagnosis: Rethinking "Nature-Deficit Disorder"', *Environmental Communication*, 7.3 (2013), 315–35; Robert Fletcher, 'Connection with Nature Is an Oxymoron: A Political Ecology of "Nature-Deficit Disorder"', *The Journal of Environmental Education*, 48.4 (2017), 226–33.

6. Tim Gill, *Urban Playground: How Child-Friendly Planning and Design Can Save Cities* (London: RIBA, 2021); Michael Martin, Andrea Jelić and Tenna Doktor Olsen Tvedebrink, 'Children's Opportunities for Play in the Built Environment: A Scoping Review', *Children's Geographies*, 21.6 (2023), 1154–70 https://doi.org/10.1080/14733285.2023.2214505.

7. Tom Wilkinson, 'Duchess's Vision Sees World's Biggest Play Park Opened', *Evening Standard*, 24 May 2023 https://www.standard.co.uk/news/uk/duchess-northumberland-b1083434.html [accessed 24 November 2023]; Amanda Hyde, '£56 for Two Hours: My Family Trip to Windsor's Extortionate New Playground', *The Telegraph*, 18 August 2023 https://www.telegraph.co.uk/travel/destinations/europe/united-kingdom/england/berkshire/windsor/family-trip-to-new-kids-playground-windsor-berkshire/ [accessed 24 November 2023].

8. National Museum, New Delhi, HR 13974/222, *Harappan Toy Cart*, twenty-fifth century BCE; Bristol Museum & Art Gallery, H1956, *Rag Ball from Grave 518, Tarkhan, Egypt*, twenty-third century BCE; The Metropolitan Museum of Art, New York, 07.286.4, *Terracotta Group of Two Girls Playing a Game Known as Ephedrismos*, late fourth–early third century BCE.

9. Kunsthistorisches Museum, Vienna, GG 1017, Pieter Bruegel, *Children's Games*, 1560.

10. Amy Orrock, 'Homo Ludens: Pieter Bruegel's Children's Games and the Humanist Educators', *Journal of Historians of Netherlandish Art*, 4.2 (2012), 1–21.

11. Nicholas Orme, *Medieval Children* (New Haven: Yale University Press, 2001).

12. Thomas Elyot, *The Book Named the Governor* (London: Dent, 1965), p. 62.

13. Richard Mulcaster, *Positions* (London: Longmans, Green & Co., 1888), pp. 114–15.

14. James Kay, *The Training of Pauper Children* (London: Poor Law Commissioners, 1838), p. 27.

15. 'Playground, n.', *Oxford English Dictionary* (Oxford: Oxford University Press, 2006).

16. Francis de Valangin, *A Treatise on Diet, or the Management of Human Life* (London: Pearch, 1768).

17. 'The Hathern Playground', *Leicester Chronicle*, 14 January 1837, p. 4.

18. 'Latest News – Mr Charles Melly', *John Bull*, 5 June 1858, p. 368.

19. Arthur Young, *A Tour in Ireland, 1776–1779*, 2nd edn (London: Cassell, 1897), p. 48; 'Boarding School for Young Gentleman at Ilford in Essex', *Morning Chronicle*, 1 August 1795, p. 5.

20. 'Open Spaces in Parliament', *The Times*, 23 February 1885, p. 4; 'Open Spaces in Hackney', *Daily News*, 18 April 1890, p. 5.

21. Luca Csepely-Knorr and Mária Klagyivik, 'From Social Spaces to Training Fields: Evolution of Design Theory of the Children's Public Sphere in Hungary in the First Half of the Twentieth Century', *Childhood in the Past*, 13.2 (2020), 93–108; Vanessa Rutherford, 'Muscles and Morals: Children's Playground Culture in Ireland, 1836–1918', in *Leisure and the Irish in the Nineteenth Century*, ed. Leeann Lane and William Murphy (Liverpool: Liverpool University Press, 2016), pp. 61–79; Essi Jouhki, 'Politics in Play: The Playground Movement as a Socio-Political Issue in Early Twentieth-Century Finland', *Paedagogica Historica* (2023), https://doi.org/10.1080/00309230.2022.21554811–21; Ann Marie Murnaghan, 'Exploring Race and Nation in Playground Propaganda in Early Twentieth Century Toronto', *International Journal of Play*, 2 (2013), 134–46.

22. Clarence E. Rainwater, *The Play Movement in the United States: A Study of Community Recreation* (Chicago: University of Chicago Press, 1922).

23. Dominick Cavallo, *Muscles and Morals: Organized Playgrounds and Urban Reform, 1880–1920* (Philadelphia: University of Pennsylvania Press, 1981); Elizabeth Gagen, 'An Example to Us All: Child Development and Identity Construction in Early 20th-Century Playgrounds', *Environment and Planning A: Economy and Space*, 32.4 (2000), 599–616; Elizabeth Gagen, 'Playing the Part: Performing Gender in America's Playgrounds', in *Children's Geographies: Playing, Living, Learning*, ed. Sarah Holloway and Gill Valentine (London: Routledge, 2000), pp. 213–29; Elizabeth Gagen, 'Landscapes of Childhood and Youth', in *A Companion to Cultural Geography*, ed. James Duncan, Nuala Johnson and Richard Schein (Oxford: Blackwell, 2004), pp. 404–19; Ocean Howell, 'Play Pays: Urban Land Politics and Playgrounds in the United States, 1900–1930', *Journal of Urban History*, 34 (2008), 961–94; Suzanne Spencer-Wood and Renee Blackburn, 'The Creation of the American Playground Movement by Reform Women, 1885–1930', *International Journal of Historical Archaeology*, 21 (2017), 937–77; Kevin G. McQueeney, 'More than Recreation: Black Parks and Playgrounds in Jim Crow New Orleans', *Louisiana History*, 60.4 (2019), 437–78.

24. Elizabeth Gagen, 'Too Good to Be True: Representing Children's Agency in the Archives of Playground Reform', *Historical Geography*, 29 (2001), 53–64; Michael Hines, '"They Do Not Know How To Play": Reformers' Expectations and Children's Realities on the First Progressive Playgrounds of Chicago', *The Journal of the History of Childhood and Youth*, 10 (2017), 206–27.

25. Valerie Wright, 'Making Their Own Fun: Children's Play in High-Rise Estates in Glasgow in the 1960s and 1970s', in *Children's Experiences of Welfare in Modern Britain*, ed. Siân Pooley and Jonathan Taylor (London: University of London Press, 2021), pp. 221–46.

26. Krista Cowman, 'Open Spaces Didn't Pay Rates: Appropriating Urban Space for Children in England after WW2', in *Städtische Öffentliche Räume: Planungen, Aneignungen, Aufstände 1945–2015 (Urban Public Spaces: Planning, Appropriation,*

Rebellions 1945–2015), ed. Christoph Bernhardt (Stuttgart: Franz Steiner Verlag, 2016), pp. 119–40; Krista Cowman, '"The Atmosphere Is Permissive and Free": The Gendering of Activism in the British Adventure Playgrounds Movement, ca. 1948–70', *Journal of Social History*, 53.1 (2019), 218–41; Roy Kozlovsky, 'Adventure Playgrounds and Postwar Reconstruction', in *Designing Modern Childhoods: History, Space, and the Material Culture of Children; An International Reader*, ed. Marta Gutman and Ning de Coninck-Smith (New Jersey: Rutgers University Press, 2007), pp. 171–90; Roy Kozlovsky, *The Architectures of Childhood: Children, Modern Architecture and Reconstruction in Postwar England* (Farnham: Ashgate, 2013).

27. Ben Highmore, 'Playgrounds and Bombsites: Postwar Britain's Ruined Landscapes', *Cultural Politics*, 9 (2013), 323–36; Lucie Glasheen, 'Bombsites, Adventure Playgrounds and the Reconstruction of London: Playing with Urban Space in Hue and Cry', *The London Journal*, 44.1 (2019), 54–74; Ian Grosvenor and Kevin Myers, '"Dirt and the Child": A Textual and Visual Exploration of Children's Physical Engagement with the Urban and the Natural World', *History of Education*, 49.4 (2020), 517–35.

28. Katy Layton-Jones, *National Review of Research Priorities for Urban Parks, Designed Landscapes, and Open Spaces: Final Report*, Research Report Series, 4 (London: English Heritage, 2014).

29. For a detailed review of the field see Laura Tisdall, 'State of the Field: The Modern History of Childhood', *History*, 107.378 (2022), 949–64.

30. Karen R. Jones, 'Green Lungs and Green Liberty: The Modern City Park and Public Health in an Urban Metabolic Landscape', *Social History of Medicine*, 35.4 (2022), 1200–1222; Peter Thorsheim, 'The Corpse in the Garden: Burial, Health, and the Environment in Nineteenth-Century London', *Environmental History*, 16 (2011), 38–68; Roy Rosenzweig and Elizabeth Blackmar, *The Park and the People: A History of Central Park* (Ithaca: Cornell University Press, 1992).

31. Bernard Mergen, 'Children and Nature in History', *Environmental History*, 8 (2003), 643–69; Simo Laakkonen, 'Asphalt Kids and the Matrix City: Reminiscences of Children's Urban Environmental History', *Urban History*, 38 (2011), 301–23.

32. James Greenhalgh, 'The New Urban Social History? Recent Theses on Urban Development and Governance in Post-War Britain', *Urban History*, 47.3 (2020), 535–45; Simon Gunn and Alastair Owens, 'Nature, Technology and the Modern City: An Introduction', *Cultural Geographies*, 13 (2006), 491–6.

33. Simon Gunn, 'The Rise and Fall of British Urban Modernism', *Journal of British Studies*, 49.4 (2010), 849–69; Otto Saumarez Smith, *Boom Cities: Architect-Planners and the Politics of Radical Urban Renewal in 1960s Britain* (Oxford: Oxford University Press, 2019); Guy Ortolano, *Thatcher's Progress: From Social Democracy to Market Liberalism through an English New Town* (Cambridge: Cambridge University Press, 2019).

34. Mathew Thomson, *Lost Freedom: The Landscape of the Child and the British Post-War Settlement* (Oxford: Oxford University Press, 2013), p. 144.

35. James Greenhalgh, *Reconstructing Modernity: Space, Power, and Governance in Mid-Twentieth Century British Cities* (Manchester: Manchester University Press, 2018).

Chapter 1

Finding space for play: 'playgrounds for poor children in populous places'

The children's playground has its roots in philanthropic, voluntary and state responses to industrialisation and urban expansion during the nineteenth century. The idea that children required dedicated play space formed part of wider efforts to ameliorate the social and environmental impacts of rapidly expanding towns and cities. However, as we shall see, there was not a simple, parallel relationship between urbanisation and the provision of public play space – playground creation also relied upon shifting conceptions of childhood, particularly the idea that children might have opportunities for play. Similarly, while there is a close association between parks and playgrounds today, in the nineteenth century the creation of public green spaces was not necessarily a good indicator of expanding playground provision. Recent scholarship on the history of public parks has tended to position the children's playground as a product of the early twentieth century, part of a wider expansion of open-air leisure amenities such as lidos and playing fields.[1] Such assumptions fail to acknowledge earlier conversations about the relationship between the interconnected ideas of public space, education and health. To address this oversight, this chapter explores the nineteenth-century experience, a crucial period that forms the background to the remainder of the book. It represents a period when ideas about the need for dedicated public spaces for children's recreation were defined, largely as an antidote to the perceived problems of urban life and the environmental consequences of industrialism, but only intermittently implemented.

The chapter builds on the work of historians and geographers interested in the imagined and material aspects of the nineteenth-century city and brings them into conversation with scholars who have examined shifting assumptions about age, education and children's play across the same period.[2] In doing so, the chapter explores the social and spatial consequences of changing attitudes towards education, exercise and urban infrastructure. It points to the intermittent and interconnected nature of philanthropic, voluntary and state action in this arena and highlights contemporary belief in the power of the urban landscape to shape the behaviour and health of individuals and communities. To make sense of these processes, the first part of the chapter focuses on the mid-nineteenth century and explores early, piecemeal attempts to create public amenities for children in Salford, Manchester and London. The second part shows how the creation of dedicated spaces for children became more widespread from the 1880s, as a remedy for the ills of the metropolitan environment and a prescription for improving the physical condition of the city's inhabitants. In doing so, it highlights a transatlantic exchange of ideas about play space provision and examines the practical debates about the ideal playground form.

Education and exercise in the mid-nineteenth century

Rapid industrialisation and urbanisation in the early nineteenth century generated significant wealth and technological innovation, but also caused major social and environmental problems. In particular, the places where the urban poor lived were increasingly seen and experienced as crowded, disorientating, dangerous and unhealthy. Publications including *Rookeries of London* (1850), *London Labour and the London Poor* (1851) and *Town Swamps and Social Bridges* (1859) all depicted a highly problematic urban environment and described the dangers for society in general and children in particular.[3] As a result, there was increasing debate about the merits of town and country, concern about the provision of spaces for recreation, and greater municipal and philanthropic efforts to tackle these problems. The playground would gradually become embroiled in these processes, but not before there was a shift in attitudes towards working-class childhood.

For many children from poor urban families, daily life was spent working and playing in the street. It had many advantages as a site for play, including proximity to home, easy sociability and space. However, in 1835 the Highway Act made street play illegal if it disrupted other traffic. The street was also increasingly imbued with negative connotations by writers such as John Ruskin, understood as a space that was both literally and symbolically dirty, diseased and dangerous.[4] At the same time, the romantic ideals of Rousseau, Wordsworth and others were increasingly influential in creating an idealised, mythical figure of the child in nature.[5] As a result, the lived experience of poor urban children was increasingly at odds with upper- and middle-class ambitions for childhood and assumptions about the city. Governments and philanthropists had long developed and implemented policies towards children, but this emerging romantic ideology began to influence public action from the middle of the nineteenth century. Legislation initially sought to limit the hours and improve the conditions in which children worked, but gradually an emphasis on a 'natural' childhood coincided with evolving ideas about education, particularly the education of young children.

The influence of progressive educational thinking on the origins of the playground can be seen in the interaction between pedagogical theory, playful practice, commerce and campaigning. In 1826, the influential German educator Friedrich Froebel had urged that 'every town should have its own common playground for the boys' so that they could learn civil and moral virtues through playing games.[6] He developed an approach to children's education that emphasised playful activities, made use of tools to support self-directed learning and included the use of materials such as bricks, sand and sawdust in the classroom. From the 1820s, British educators including Samuel Wilderspin and Robert Owen also promoted the place of play in children's education, as part of a wider infant education movement that positioned schooling as a solution to criminality.[7] An engraving of the playground at the Home and Colonial infant school in London shows a generously equipped, enclosed outdoor space with a variety of apparatus including see-saws, climbing ropes and bars (Figure 1.1). The playground appears to have been a key component of an outdoor, physically energetic education for young children. In addition, and significantly for the subsequent story of the public playground, both Wilderspin and Froebel combined their pedagogical theory and teaching practice with a commercial sideline supplying educational equipment, underlining the enduring connection between commerce and play.

For many in the infant education movement, including Wilderspin, it was the plight of children from poor families in particular that most needed the corrective influence of infant education. But while the school day helped to remove some poor children from the workplace, it did little to tackle the perceived problems of playing in the city street.[8] For Wilderspin's friend Charles Dickens it was the unsavoury streets that were particularly problematic. Writing in *Household Words* in 1850, Dickens described how poor children were 'generally born in dark alleys and back courts, their playground has been the streets, where the wits of many

GYMNASIA AND PLAY-GROUND OF THE CHILDREN OF THE HOME AND COLONIAL INFANT SCHOOL SOCIETY, GRAY'S-INN-LANE.

Figure 1.1: Gymnasia and playground of the Home and Colonial Infant School, London, wood engraving, c.1840, Wellcome Collection. Public Domain Mark.

have been prematurely sharpened at the expense of any morals they might have'.[9] But he did more than write about these problems and, although overlooked in otherwise comprehensive biographies of his life, Dickens also sought to provide more salubrious places for poor children to play.

In January 1858, Dickens and the reforming politician Lord Shaftesbury launched the Playground and Recreation Society in an effort to create 'playgrounds for poor children in populous places', away from the 'variety of temptations' and 'bodily evils' to be found in the street.[10] Later that year, a deputation from the Society that included Dickens and park advocate Robert Slaney MP met with government ministers to promote the cause.[11] No doubt partly as a result of Dickens's reputation, the Society received considerable publicity, including support from Henry Mayhew's satirical magazine *Punch*. While it suggested that 'ragged playgrounds' would remove the annoyance of children playing on the street, Dickens emphasised that play spaces would benefit children's physical and mental health and consequently the strength and status of the nation.[12] These benefits were repeated in parliament, where Slaney promoted a bill that would enable the provision of public spaces for recreation.

The subsequent Recreation Grounds Act (1859) specifically permitted the creation of playgrounds for children.[13]

Despite Slaney's work in parliament and Dickens's high-profile involvement in the campaign, the Playground and Recreation Society was short-lived. By May 1860, *The Times* reported that it had 'lately died a natural death, obviously from the impossibility of creating spaces, or providing funds adequate to the enormous cost of purchasing ground in the metropolis'.[14] While the high price of land and lack of funding undoubtedly presented problems for Dickens, these were issues which later campaigners were able to overcome. In mid-century London, it seems likely that wider society was not yet ready to embrace the children's playground as a solution to the problems of the city. Neither celebrity endorsement, the nuisance and danger of street play, nor the future potential of a healthier working-class childhood were convincing enough to attract state support, philanthropic funding or the allocation of dedicated public spaces for children. Instead, the earliest spaces set aside specifically for children were seen in another response to the problems of the nineteenth-century city, the campaign to create public parks. Even here, however, the playground did not initially feature prominently.

The city park had long been understood as a space of health and recreation. But while there is a strong association today between the park and the children's playground, they were not intimately connected from the outset. As a response to the problems of the industrial city, green space advocates envisaged the public park as a way to provide fresh air, a dose of nature, gentle recreation and cultural enrichment. Parks were imagined as green lungs, part of an urban respiratory system which supported the healthy functioning of the wider city. In adopting such bodily metaphor, park advocates imagined that green spaces would help to ventilate overcrowded streets, circulate clean air and disperse noxious miasmas.[15] In addition to this ostensibly biological function, the park could also provide a cure for the social and moral problems of the city. The 1833 Select Committee on Public Walks endorsed green spaces as a solution to the ill health, poor hygiene and intemperance of working-class city dwellers. Motivated by civic pride and philanthropic charity, later proponents also saw them as a vehicle for educating and enriching the lives of the urban poor. By imagining the city as a living organism, whose physical and moral ills could be cured through medico-environmental interventions, early park advocates provided an important conceptual framework for later proponents of the children's playground. Adapting the urban environment in this way would help to reshape the health and behaviour of its working-class population.

While the common refrain of 'parks for the people' may have implied a democratic purpose, green spaces were often shaped by gender- and class-based values, which invariably stressed purposeful and segregated forms of rational recreation, rather than energetic exercise. At People's Park in Halifax, early by-laws prohibited dancing and games, while at Longton Park in Stoke-on-Trent, bicycles, tricycles and dogs were banned and the park superintendent advised against installing facilities for children or sport.[16] Instead, gentle perambulation would allow visitors to interact with and learn from an ordered version of nature while fresh air would ward off disease, and the bandstand or tearoom provided an appropriately salubrious break from daily routines. Significantly for this story, parks were also shaped by assumptions about age. Children were expected to imitate adult behaviours, strolling on the paths or admiring fauna and flora from a distance. At a minority of green spaces, including Saltaire Park in Bradford and the Brewer's Garden in Stepney, young people were barred from entering altogether.[17]

Instead, designers took their cues from English landscape parks, with picturesque lakes, groups of trees, expanses of grassland and serpentine walks. Many early public parks were created in this vision; the plans for Victoria Park (1845) and Birkenhead Park (1847) were both an expression of these aesthetics, as was an unrealised plan for an Albert Park in north London.[18] Furthermore, while advocates imagined the park as a remedial space of natural beauty and healthy recreation, in practice there was considerable continuity in earlier patterns of behaviour. Regulation and supervision sought to manage the way that visitors used parks, but earlier activities such as picking flowers, intimacy between couples and most likely children's play all challenged attempts to impose alternative, park-appropriate values and uses.[19] With evidence that there was considerable consensus within the park community around what constituted 'respectable' behaviour, it is doubtful that the park keeper would be the only adult attempting to moderate children's instinctive playfulness.[20] At the same time, the early appropriation of the ornamental lakes in Victoria Park as a site for bathing, often by large numbers of children, highlights how the creation of norms relating to the use of such spaces was a negotiated process, one where designers' intentions and administrators' expectations were modified in practice by park users.

In one part of the country there was an apparent exception to the marginalisation of children in park landscapes. Manchester was an archetypal 'shock city' of the nineteenth century, where industrialisation created economic growth, but also resulted in social and environmental problems, particularly working-class poverty and ill health.[21] For a local curate, William Marriott, children in particular suffered 'the pain, the sickness, the filth, the disease, and the thousand gross immoralities, and brutish vices and degrading crimes' that resulted from such conditions.[22] For local reformers, the creation of new public parks was central to efforts to mitigate the physical and moral consequences. Largely paid for by local subscription, Peel Park in Salford and Queen's Park and Philips Park in Manchester were opened to the public on the same day in 1846. Like contemporary parks elsewhere, they were designed in the landscape style, reinforced normative gender values and provided open space for the working-class to take part in moderate exercise and interact with nature. Unusually, the parks also included specific facilities for children, designed to encourage energetic exercise.

A plan of Peel Park from 1850 shows that space for children was provided in addition to an archery ground, skittles ground and gymnasium.[23] A rectangular 'Girls Play Ground' and a circular clearing for a girl's swing were tucked away in the shrubbery on the park boundary, while a boy's swing was positioned close to the quoits ground, again hidden among the planting. In Philips Park, the 'play grounds' were laid out with a gravel surface and bordered by an earth bank planted with privet hedging.[24] Contemporary accounts of the three parks described how the girls' playgrounds provided space for skipping and shuttlecock, while the gymnasium provided boys and men with equipment for athletic exercise. Laid out with advice from the local Athenaeum Gymnastics Club, apparatus included a 7m frame which supported climbing ladders, poles, bars and ropes, a vaulting horse and a giant stride.[25] The latter, also known as flying steps, was a tall, upright pole with a revolving top on which ropes were attached that allowed users to take giant steps around a circle (see Figure 1.2 for a later example). For its advocates, it provided 'a most useful article in the muscular education' and made the gymnast appear to be 'endowed with wings'.[26] In common with the wider provision of recreational facilities in parks, these were not facilities for instinctive and unstructured play that might otherwise have taken place in the street. Instead, the equipment represented an attempt to provide for rational exercise by children.

The new parks sought to civilise the natural world and children's play within it, but this was not an unproblematic task. The inclusion of engineered gymnastic structures, for example, might seem at odds with the typical approach to park design that was grounded in a pastoral landscape aesthetic. However, industrial technology had long been a feature of green spaces. It underpinned the automata that featured in the parks and gardens of the European aristocracy, as well as the operation of fountains and construction of follies.[27] Similarly, Victorian public parks and the people that used them were invariably influenced by industrial

Figure 1.2: Giant's stride, Bayliss Jones and Bayliss Ltd, 1912, National Archives, WORKS/16/1705.

architecture, materials and technology. Iron, for example, was an essential material used to create seemingly bucolic green spaces and shape their use, an essential component of bridges, bandstands, railings and glasshouses.[28] Early gymnastic apparatus for children was often similarly constructed from iron, and as we shall see in later chapters, this use of robust metals would inform both visions of the ideal playground and the companies that could supply such equipment.

More problematic was the attempt to combine bucolic landscapes with recreational opportunities for children. Joshua Major, the designer of the Salford and Manchester parks, resolved this issue by prioritising the former, recommending caution in providing facilities for recreation and particular care in siting them. He emphasised that aesthetics should take priority over practical amenities such as playgrounds, arguing that such features should never 'interfere with the composition and beauty of the general landscape'.[29] Playgrounds that had initially been positioned in the centre of Philips Park were removed and consolidated on the boundary, so that the 'unrestrained merriment of the factory girls who used the swings' would no longer impinge on the view.[30] As well as being secondary to aesthetic concerns, the provision of space for recreation could be trumped by economic considerations. In 1850, Peel Park administrators suggested closing the playgrounds for part of the year to preserve the quality of the grass.[31] Protecting the turf in this way would enable more grass to be harvested for hay and then sold, raising around £26,630 in today's money (£30 in 1850) to help offset park maintenance costs.[32]

Despite provisions in the 1859 Recreation Ground Act which legally permitted public authorities to create playgrounds for children, few did so. Joshua Major's other notable landscape designs seem to have included no dedicated facilities for children.[33] Elsewhere, plans for Sefton Park (1872) in Liverpool, Finsbury Park (1869) in London and Roundhay Park (1872) in Leeds did not initially include child-specific amenities. Stamford Park (1880) in Altrincham was unusual in providing a boys' playground and girls' playground, but again they were hidden from view on the park boundary, enclosed by trees and shrubs.[34] At the opening of Victoria Park in Portsmouth in 1878, one commentator felt that the otherwise undesirable railway line, which divided the new space in two, did perform a useful function, separating the giant stride and spaces for recreation from the rest of the ornamental park landscape.[35]

As such, children were rarely a primary constituency in the mid-century park community and dedicated spaces for children were not common. When amenities for children were provided, they were invariably located on the marginal boundaries of the park, hidden from view and subservient to the wider landscape aesthetic, economic considerations and expectations of appropriate park behaviour. The children's playground was not yet a defining characteristic of the public park.

Just as tentative steps were being taken to provide spaces for children in some towns and cities, there were also attempts to promote more energetic forms of activity in parks. For landscape historians, these interconnected processes have resulted in some uncertainty about whether amenities were intended for children or adults, particularly the provision of gymnasiums in public spaces. The garden historian Susan Lasdun has contended that the gymnasium installed in 1848 at Primrose Hill in London was the first children's playground.[36] Although the evidence from Manchester and Salford suggests this is an unsound assertion, it is a claim worth exploring in more detail as it points to an important influence on the form of the playground, one which focused less on the perceived benefits of green space and instead emphasised physical exertion. There had been attempts to provide space for open-air athletic exercises earlier in the century. In 1825 the German 'professor of gymnastics' Karl Voelker began offering lessons for military gentlemen close to Regent's Park in London.[37] A sketch of his gymnasium showed a range of apparatus including bars, ladders and poles, all positioned in an outdoor area enclosed by a high brick wall.[38] By 1827, a cartoon published in

Lady's Magazine parodied the gymnasium with its high and giddy mast, risky javelin throwing and other exercises that seemed to provide as much amusement to onlookers as they did health to participants.[39] Perhaps partially as a result, attempts to increase participation in gymnastics were not wholly successful and Voelker's London Gymnastic Institution closed in 1827 due to lack of income.[40]

Despite the cynicism and setback, the perceived benefits of gymnastic exercise did become more widely acknowledged. Around the same time that Voelker was working in London, a Swiss professor of gymnastics, Peter Clias, organised gymnastic courses at military and naval colleges across Britain and published a general introduction to athletic exercise.[41] Donald Walker's *British Manly Exercises* promoted a similar approach and ran to ten editions between 1834 and 1860.[42] As part of this mid-century enthusiasm for energetic exercise, an outdoor gymnasium was installed by the Commissioners of Woods and Forests at Primrose Hill (as Lasdun noted). Much like Voelker's earlier enterprise, it included gymnastic apparatus such as ropes for swinging on, poles for climbing up, horizontal and parallel bars. Newspaper accounts suggested that exercising on the equipment would provide new vigour, improved health and a strengthening of the mind, while also making clear that this was not a space specifically for children. An engraving in the *Illustrated London News* provides an insight into the way the gymnasium was both perceived and represented to the public.[43] Adults are seen exercising and spectating, many dressed as might be expected of the gentlemen that Voelker had earlier hoped to attract to his gymnastic lessons. Older boys are shown climbing, another playing with a hoop, but the size of the apparatus would have made it difficult for many younger children to use the gymnasium, while women and girls are relegated to the sidelines as spectators. The most notable child, in the foreground of the image, conforms to contemporary representations of working-class children, portrayed as a costermonger or street seller straight from the pages of Henry Mayhew's *London Labour and the London Poor*. For working-class children at least, the gymnasium seems to have been a site of work rather than play.

Providing public amenities for active recreation was far from straightforward. In 1863, regulations were introduced to manage demand for the equipment at Primrose Hill, limiting the length of time each piece could be used and noting that abusive language or wilful damage would result in exclusion or prosecution.[44] At the same time, there was nothing inherently respectable about those responsible for supervising the gymnasium. *The Standard* reported that it had become both 'a very disorderly place' and a site of unscrupulous administration, with park constables demanding bribes before people could use the equipment.[45] As such, this was not a space for playing freely as an end in itself. Neither the regulations nor newspaper accounts specifically mention children or play, and it seems highly unlikely that park staff, gymnasts or the wider public would have seen this as a space exclusively for children. But in creating a public, open-air facility for energetic exercise, the example at Primrose Hill serves as a useful reference for the justification and design of later spaces that were set aside for children in particular.

In summary, between the 1840s and 1870s there were only sporadic and localised attempts to create dedicated public spaces for children's recreation, although many of the factors that would influence later advocacy had their roots in this period. Mid-century investigations had highlighted the deleterious effects of the urban environment on its poor inhabitants generally and children especially, while education, restorative exercise and interaction with nature were positioned as potential solutions. The vision of a universal, natural childhood contrasted sharply with the perceived reality of the poor urban child, in an overcrowded home and with nowhere to play except the street. However, attempts to provide playground spaces as a solution were either unsuccessful or at odds with the dominant landscape aesthetic of the public park movement. This would change in the 1880s as shifting conceptions of childhood stabilised and heightened anxiety about the consequences of urbanisation demanded more pragmatic solutions.

Childhood and urban anxieties in the late nineteenth century

If attempts to create spaces for play across the middle of the century had achieved decidedly mixed results, by the 1880s heightened concern about the problematic urban environment and changing conceptions of childhood meant that conditions were more conducive to playground creation. London in particular became a focus of concern, and the playground was increasingly promoted as a solution to at least some of the problems facing the capital. Utopian visions of a healthy city, pragmatic attempts to improve the housing of the poor and the work of open space advocates all promoted dedicated spaces for children. At the same time, a series of legislative interventions created 'spare time' for children beyond the school day. The 1880 Education Act made school attendance compulsory between the ages of five and ten, increasing to twelve in 1899, while the 1891 Elementary Education Act effectively made schooling free.[46] At the same time, earlier park rhetoric, particularly ideas about the benefits of fresh air and interaction with nature, and the possibilities of energetic gymnastic exercise were brought together in the playground for significant political and social purposes.

The depiction of poverty-stricken working-class childhood by sensationalist journalism and more sober social science provided a particularly persuasive impetus to late nineteenth-century urban reform. In *The Bitter Cry of Outcast London* (1883), congregational minister Andrew Mearns lamented that 'the child-misery that one beholds is the most heart-rending and appalling element', a tragedy made worse because 'many of them have never seen a green field'.[47] William Booth, founder of the Salvation Army, visualised millions of poor urban children enduring a 'miserable subsistence', their 'amusement curtailed to the running gutter'.[48] These findings contrasted sharply with romantic notions of childhood, a life stage increasingly imagined as a time of natural progress, education and hope for the future. This idealised childhood was a long way from the perceived reality of most urban working-class children, providing an important motivation for many reformers and philanthropists in the 1880s.

For some, occupying children's newly created leisure time with appropriate activities was the priority. The driving force behind such efforts was the novelist and social worker Mary Ward. Built upon accounts of the slums and descriptions of poverty, but mainly concerned with the behaviour of poor children, the organisers of the Children's Happy Evening Association (1888), various Guilds of Play and other evening and holiday play schemes all attempted to divert children from the dangers of the street by occupying their leisure time. Supervised activities included drill and dancing, lantern talks and basketwork and generally took place inside. The Happy Evening Association, for example, made use of school premises to provide constructive, supervised play opportunities for young children outside of school hours, while the Guilds of Play focused particularly on dancing for girls.[49] However, the schemes were only available one or two nights a week or during the school holidays and were primarily concerned with occupying children's leisure time, rather than adapting the urban fabric to mitigate the impacts of the city on children's health.

In contrast, there were philanthropic reformers who were more concerned about the spaces where public play should take place. In 1866, the housing reformer Octavia Hill described clearing some old stables to create a playground for poor girls at one of her model housing schemes at Freshwater Place, Marylebone. Fenced, gravelled, planted with small trees and equipped with swings, by 1869 'the playground never looked so pretty', she wrote.[50] As a member of the Kyrle Society, Hill also promoted the principle of the playground, stating in 1883 that 'children want playgrounds' and that when provided they 'would not be obliged to play in alleys and in the street, learning their lessons of evil, in great danger of accident'.[51] Beyond Freshwater Place and Hill's advocacy, the playground also appeared in more radical visions of the future. In 1876, the noted sanitarian Benjamin Ward Richardson reimagined the city as a space of health, hygiene and cleanliness. In his highly detailed description of *Hygeia: A City of Health*, garden squares at the back of working-class housing would be 'ornamented with flowers and trees and furnished with playgrounds for children'.[52] Richardson's writing and Hill's advocacy both helped to ensure that the playground was firmly planted within both practical action and utopian visions for a more humane urban environment.

At the same time, free-to-use public play spaces were not the inevitable nor only response to the problem either. An 1873 sketch of Victoria Park in the *Illustrated London News* included expected park features such as the pagoda, cascade, lake and boat house, as well as a small detail showing the 'swings and roundabouts'.[53] But rather than a children's gymnasium, the detail appears to show a covered carousel and swing boats, both more commonly associated with the fairground rather than contemporary ideas of the playground. A later London County Council (LCC) publication confirms that these playful features were only available on payment of a fee. Alongside regulations relating to the use of gymnasiums and other park amenities, the 1894 park by-laws included a separate prescription that 'the charge for the use of swings erected by private persons in parks or on open spaces is to be one penny per person for five minutes'.[54] The provision of private swings was presumably common enough to warrant a bylaw being created and significant enough that the council sought to control the cost. Ten years later, the revised LCC by-laws included no mention of privately operated swings. The influence of the fairground and amusement park on the form and function of the children's playground are explored in more detail in the next chapter, but one reason for the shift from private to public swings was the advocacy of philanthropic organisations who promoted free-to-access and publicly maintained spaces for children's recreation.

In the early 1880s Dickens's unsuccessful Playground Society received some nostalgic publicity, but it was the formation of the Metropolitan Public Gardens, Boulevard and Playground Association in 1882 which marked the beginning of more determined efforts to provide poor children with dedicated space for recreation.[55] Despite shortening its name to the Metropolitan Public Gardens Association (MPGA) in 1885, it remained concerned with the provision of spaces for children. The Association was founded by the aristocratic philanthropist and ardent imperialist Reginald Brabazon (Earl of Meath, Lord Chaworth, 1841–1929) and the surgeon and journalist Ernest Hart (1835–98).[56] Brabazon had been a member of the Kyrle Society but became increasingly keen to focus more specifically on the practical creation of public spaces in poor neighbourhoods, in contrast to the somewhat abstract aesthetic ambitions of the Society. Ernest Hart was chair of the National Health Society, editor of the *British Medical Journal*, and his involvement in the MPGA was part of his wider interest in social, medical and environmental issues. This combination proved to be particularly significant in the story of the playground in that it brought together anxieties about the environmental problems of the city, medico-moral understandings of working-class poverty, public park rhetoric and concern about the status of empire – and presented the public garden and gymnasium as a pragmatic solution to all of these concerns.[57] Unlike Benjamin Ward Richardson's imagined city of health, the MPGA did not promote utopian visions and instead pursued more modest responses to the problems of the city. In doing so, the MPGA's cause resonated with the politics and anxieties of the capital's upper- and middle-classes, particularly the widely held concerns about social, biological, environmental and national degeneration.

Theories of degeneration grafted Darwinian ideas about evolution onto the city and its population, asserting that urban environmental maladies were leading directly to hereditary health problems and the social and biological regression of the nation. For Brabazon the 'smoky and grimy city' led directly to 'pale faces, stunted figures, debilitated forms, narrow chests, and all the outward signs of a low vital power' among the working-class population.[58] However, he was less interested in the consequences for individuals and more concerned that a weak and unhealthy working population would threaten the nation's place in the world. Brabazon's commitment to Empire saw him play a leading role in nearly every campaign to promote the imperial cause to children, including the fabricated tradition of Empire Day (1904).[59] He felt that open spaces could promote the health and subservience of the working-class, with beneficial consequences for Britain's military, commercial and imperial status. This association between the urban environment, public health and the status of the empire held widespread appeal. For example, the president of the Manchester Medical Society, William Coates, emphasised the role of the medical profession in tackling national degeneration. Echoing Froebel's assertion from a century earlier, he argued that 'public gymnasiums should be provided by the municipality in all large towns'.[60] As a result, the wellbeing of working-class children became a significant feature of medical and environmental debate. For one commentator, living in overcrowded conditions and playing in the street was resulting in the evolution of 'strange creatures called the children of the town'.[61] For the MPGA, such children could be 'healthy neither mentally, morally, nor physically'.[62]

In speaking to such widespread concerns, the MPGA was able to attract members, donations and influence. Supporters included royalty, landed gentry, politicians, writers, physicians, scientists and clergy, and the organisation swiftly developed significant political leverage and accumulated considerable financial resources.[63] The MPGA's rapid prominence has attracted historical analysis but the scholarship has mainly focused on the Association's efforts to create urban gardens for the working classes in general. However, from the outset the Association's objectives included the provision of spaces specifically for children. Its second annual report made it clear that they were seeking to create three types of open space: gardens, playgrounds and gardens with playgrounds. Gardens would provide spaces of respite from the city, principally adult resting places with benches, grass, flowers, shrubs and trees. In contrast, playgrounds were intended for the exclusive use of children, a site for simple gymnastics. When garden and playground were combined, they would be laid out with 'broad stretches of concrete pavement, interspersed with shrubs, and trees, and grass, and seats'.[64]

In promoting public gardens as a response to the problems of urban childhood, the MPGA's rhetoric demonstrated considerable continuity with the ideas of earlier park advocates and their attempts to bring nature into the city. Urban green space was still understood as a restorative, educational and refining tonic. Ernest Hart wrote of the 'alchemy of nature' and its ability to tackle the evils of the urban environment.[65] The MPGA also continued to associate open space with the biological wellbeing of the wider city organism, with new public gardens improving not just the local area but also providing the whole metropolis with 'much more general lung power'.[66] In common with parks elsewhere, nature also needed a degree of protection and images of gardens created by the MPGA show railings encircling lawns and flower beds, presumably in an effort to protect tender aspects of nature from the threat of destruction posed by both children and adults.[67]

In a modification to earlier park values, the MPGA argued that proximity and scale were key issues in open space provision and that earlier public parks had not delivered their anticipated benefits widely enough. Although large parks may have received thousands of visitors each day (a Whit Monday census in Victoria Park counted over three hundred thousand), these numbers seemed inconsequential when compared to the many more who lived in overcrowded neighbourhoods, unable to afford transport around the city.[68] Such criticism was not new in the 1880s. As early as 1861 the lawyer and later MP William Marriot asserted that public parks were 'too far apart to supply the lungs which a town like Manchester requires' and too far away from children's homes to provide a useful place for play.[69] Such assertions were repeated more consistently from the 1880s. Walter Besant, author and MPGA member, felt that for 'the children and the old people ... of that vast region which lies north of the old London wall – a densely populated district inhabited almost entirely by the working classes – London might almost as well be without any parks at all'.[70] For local councillor Reginald Bray, such far-off parks 'containing soot-stained grass and a few dishevelled sparrows' and lawns 'on which no one must tread' were no longer fit for purpose.[71] For the MPGA and its supporters, the large, mid-century public parks that had been located on the edge of expanding cities had not delivered the benefits that had been expected.

As a result, rather than expect people to travel to large landscape parks, the MPGA sought to create smaller spaces within working-class neighbourhoods. Writing in 1887, Brabazon stated that 'however important it may be to provide a few large and expensive parks for the people, it is of still greater importance to create small gardens and resting places within easy distance of their homes', echoing Dickens's earlier ambition to provide a daily source of healthy recreation.[72] By 1893, Brabazon had refined this idea further, emphasising the need for smaller, equipped and segregated spaces. He called for the creation of 'a children's playground divided into two portions, one for boys and one for girls, both supplied with gymnastic apparatus' within a quarter of a mile of every working-class home.[73] This adaptation in the scale and siting of public green spaces was accompanied by a shift in ideas about their potential role in promoting healthiness, particularly the type of exercise that was best suited to tackling the problems of urban degeneration. Brabazon and the MPGA emphasised the importance of energetic physical activity rather than more genteel forms of public recreation, while also placing the provision of specific facilities for children more firmly within the park boundary. This emphasis on vigorous exercise had its roots in interconnected fields of thought and practical action, including the introduction of physical education in schools, high-profile debate about the merits of different systems of exercise, the respectability of physical exercise in public and the perceived shortcomings of military recruits.

The inadequacy of British troops in the South African War (1899–1902) is often cited as the main impetus for the provision of physical training facilities for adults in outdoor public spaces. But, while military setbacks may have contributed to turn-of-the-century anxiety about the status of the empire, there had long been concern about the physical strength of both military personnel and children, as well as competing theories about how to best address such concerns. The apparatus-based gymnastics promoted by Voelker and Clias gradually gained greater currency, so that by the late 1860s such exercise was a regular part of military training.[74] At the same time, a competing system based on the theories of Pehr Henrik Ling from Sweden promoted 'medical gymnastics' performed largely without apparatus. Promoted energetically in Britain by the physician Mathias Roth, Ling-inspired gymnastics could seemingly provide both muscular strength and more general health benefits including better posture, improved deportment and even relief from chronic disease.[75] The simple exercises and negligible equipment costs made it a favourable choice for the physical education of working-class children in state-funded elementary schools.

These two approaches to gymnastic exercise were played out in the rhetoric and practical work of the MPGA, which incorporated aspects of both. By the 1890s, Brabazon was increasingly convinced that physical training for working-class children in particular was of vital importance. He had previously worked in Germany at a time when advocates of apparatus-based gymnastics were publicly and enthusiastically promoting their ideas, including the use of the wooden vaulting horse and balance beams, along with structures made from scaffolding, ladders, poles and ropes.[76] At the same time, the MPGA invested gymnastic exercise with advantages that went well beyond muscular development, subscribing to the wider sociomedical benefits that were associated with Ling, even paying for an instructor in Swedish gymnastics for the London School Board.[77] This loosely defined association between gymnastic exercise and wider physical, medical and moral health echoed the correlation between parks as lungs and the wider healthy functioning of the city. When combined with the provision of amenities for children, they created a similarly evocative and malleable concept that would unite a broad range of constituents behind the principle of playgrounds for children.

Although the MPGA's design ambitions were strongly shaped by leading members, the Association also employed landscape designers to apply and adapt these principles to individual sites and circumstances. At a time when the landscape sector was a male-dominated profession, the MPGA was unusual in employing female designers to lead its practical work. Fanny Wilkinson (1855–1951) had taken the unusual step of studying at the male-dominated Crystal Palace School of Landscape Gardening and Practical Horticulture, before starting work for the MPGA. Between 1884 and 1904 she designed and supervised the creation of over seventy-five open spaces for them, as well as planning the layout of Vauxhall Park for the Kyrle Society.[78] Wilkinson resigned in 1904 to become the first female principal of the Swanley Horticultural College and Madeline Agar replaced her as the MPGA's designer. Agar would go on to work for the MPGA for almost twenty-five years and published advice on the design of domestic gardens.[79]

In appointing female designers, the MPGA might seem like a progressive organisation. In reality, the appointment of Wilkinson and Agar highlights the conservative values that shaped the place of women in society and their apparent suitability for certain tasks and roles. In a House of Lords debate, Brabazon asserted that a ban on women being elected to the LCC should be lifted, not because equality was important, but rather because he felt women were better suited than men to dealing with aspects of the institution's work. He stated that women's natural 'aptitude for details' made them best placed to oversee the council's care of children, 'the housing of the working-classes, matters relating to the wellbeing of the poorer classes, and social reforms generally'.[80] Hardly a radical suffragist, Brabazon expounded conservative social norms that linked women's apparently inherent biological qualities with spheres of social, economic and environmental work. Alongside responsibility for raising children, domestic gardening in particular was seen as an appropriate activity for middle-class women. This presumably made the design of gardens for children a particularly suitable task, especially as playgrounds were understood in part as a remedy for the inadequacy of working-class homes. At the same time, Wilkinson and Agar forged highly influential careers, working on the design and construction of high-profile public spaces at a time when there were few women in the landscape profession. As such, they played a significant role in shaping the form of spaces set aside for children's recreation in the late nineteenth and early twentieth century.

Despite this considerable continuity in designers and rhetoric, there was not a single dominant vision for the ideal form of a children's playground, and three approaches were in circulation around the turn of the century. Firstly, the 'levelled and gravelled' ground for play, seen in Philips Park and Freshwater Place, remained an influential conception. Secondly, the outdoor gymnasium was increasingly associated specifically with children's public recreation as a result of the MPGA's high-profile advocacy. Thirdly, the comprehensive playgrounds that typified provision in the USA achieved notable publicity in Britain at this time. The next section explores these competing visions in more detail.

Despite the emphasis on gardens and greenery, the MPGA continued to create sombre playground spaces that resembled those from the middle of the century. In the 1840s and 1850s, fashioning a recreational space for either adults or children invariably involved creating a flat, level site and covering it with gravel. Letters between health boards, vestries, schools and central government often referred to works to level and gravel an area to make it suitable for recreation.[81] In some instances, the MPGA continued this approach. Spa Fields in Clerkenwell had previously been a tea house and pleasure garden (1770) and then a burial ground (1777), notorious in the 1840s for its 'pestilential condition' and illicit exhumations.[82] In 1885, the Association drained the site and imported a large amount of shingle to specifically create a ground for children, part of a wider process that saw over one hundred disused graveyards reclaimed as public spaces.[83] At Spa Fields, the playground was primarily an open area for recreation, much like a parade ground provided an obstacle-free space for marching. Spa Fields in particular performed this dual role; once levelled and gravelled it provided a space for the 21st Middlesex Rifle Volunteer Corp to drill and only at other times was it somewhere that children could use.[84]

Elsewhere, Little Dorrit's Playground (1902) was similarly described as a levelled area specifically intended for use by children. It cost £2.4m (£5,600) to purchase the land and £196,000 (£450) to level, gravel and drain it.[85] The *Illustrated London News* shared a sketch of the newly opened space (Figure 1.3), which provides an insight into the way that this type of playground, its aesthetics and its use were presented to the wider public. It was shown as a space to play active games, use outdoor toys and socialise, all activities that would have previously taken place in the street. It also represents gender- and age-specific ideas about the way that children and adults would behave in such a space. Girls are shown talking in pairs, carrying a small baby and playing with toys. Boys are shown pretending to be harnessed horses and a driver, galloping around the playground, while adults are seen in the background, close to the surrounding blocks of flats. Similar spaces and behavioural expectations were created beyond London too. In Cardiff, Loudoun Square was partially converted into a level playground and covered with gravel at a cost of over £110,000 (£257).[86] Such spaces seem to have been primarily designed as substitutes for the street, sites where children could continue with existing play habits, away from dirt and danger.

At the same time, the MPGA used the term playground to describe a second, quite different spatial form. In 1884, it leased half of the old Horsemonger Lane Gaol site and converted it into a children's playground, at a cost of £200,000 (£365). But rather than simply a levelled area set aside for children, the new Newington Recreation Ground also included apparatus. The playground was divided in two by low fencing, creating one part for girls with swings and see-saws and one part for boys

Figure 1.3: Little Dorrit's playground by H. Seppings Wright in the *Illustrated London News*, 8 February 1902, p. 208, © Illustrated London News Ltd/Mary Evans.

with gymnastic apparatus and giant strides.[87] This example typifies the MPGA's approach to the acquisition, design and maintenance of public spaces. They secured access to the site, in this instance by leasing it from the landowner, designed the new playground and arranged for it to be laid out, and then sought to place responsibility for maintenance on the local vestry.[88] It followed a similar process in Islington, north London, when it helped to create the children's gymnasium at Norfolk Square Playground.[89] This approach saw the provision of children's gymnasiums expand considerably. Myatt's Fields, laid out by Wilkinson and the MPGA, included two children's gymnasiums, one for boys and one for girls. Finsbury Park included gymnasiums for both children and adults and a carpenter's workshop to undertake repairs, while in Battersea Park a children's gymnasium was planned to compliment the adult equivalent.[90] Records from 1889 describe Victoria Park without specific facilities for children, but by 1892 an update from the superintendent reported that the new children's gymnasium was in use.[91]

A third vision for the playground asserted influence from across the Atlantic. But rather than simply a levelled and gravelled substitute for the street or an equipped children's gymnasium, in the USA it was the 'organised playground' that increasingly dominated conceptions of recreational provision for poor city children. There had been links between US and

British park advocates and designers since at least the 1840s, epitomised by Frederick Law Olmsted's visit to Birkenhead Park and its influence on the design of Central Park (1858) in New York. In 1880, Olmsted argued that smaller parks located at regular intervals would be more effective than larger green spaces, and Brabazon later promoted this idea to the LCC parks committee and shared details of his conference attendances and park visits in the USA.[92] Henry Curtis, founding member of the Playground Association of America, reported in his 1915 book, *Education Through Play*, on the ideas and actions of Brabazon and the MPGA.[93] There was also an exchange of information and ideas at a governmental level too.

In 1917, the British Ministry of Reconstruction asked their ambassador in Washington to find out more about the playground movement in the USA, although it is not clear from the records what motivated the request. The State Department's comprehensive response provided both a detailed bibliography and a range of pamphlets describing playgrounds across the country.[94] The communiqué highlighted the extent to which the playground movement in the USA had relatively quickly shifted from a primary concern with ameliorating the physical conditions of children in the urban environment, to a wider notion of reforming the child, through appropriate physical, moral, spiritual and nationalistic instruction. Sand gardens and small parks were replaced in the early twentieth century by a comprehensive community service that included structured educational activities, including gardening, debating and sewing. As such, the US playground increasingly resembled a formal educational establishment whose purpose was to teach both young and old about progressive civic hygiene.[95] This was seen in Olmsted's 1891 design for the Charlestown Playground in Boston, which included a large open area for organised activities, avenues of trees on the perimeter and gymnastic equipment located on the southern boundary.[96] The US playground incorporated the levelled and gravelled space for games, with the apparatus of the children's gymnasium and the structured activities promoted in Britain by the Children's Happy Evening Association and others. The resulting 'organised playground' received wider publicity in the UK, largely due to the work of play scheme advocate Mary Ward.[97]

These different visions for the playground – levelled and gravelled, equipped and organised – were all in circulation in the early twentieth century. While the latter did not become common in Britain, the idea that children needed guidance in their play did resonate with some. Contemporary theories about play conceptualised it as an activity where children either spent surplus energy or recuperated lost vigour, practised inherent survival skills or took steps in a journey from individual savagery to civilisation.[98] There seems to be little explicit reference to these notions of play in the rhetoric of the MPGA, concerned as it was with the redemptive possibility of nature and exercise, rather than necessarily with children's instinctively playful behaviour. However, in the eyes of some observers, many poor urban children did not know how to properly make use of playgrounds. For one commentator, 'the poor little creatures sit or stand listlessly about, idle and bewildered, not knowing what to do, not knowing how to play'.[99] While this points most significantly to the disconnect between the reformers' ideas about childhood recreation and children's instinctive, playful preferences, for some playground advocates it meant that children needed to be taught how to play if they were to become model citizens of the future.

As Carole O'Reilly has shown, public parks played a role in promoting active urban citizenship and a sense of communal responsibility. In spaces such as Heaton Park in Manchester, Victorian moralism was gradually replaced by Edwardian pragmatism, with parks increasingly imagined as spaces of shared social responsibility for health, where individual exercise could contribute to a collective, colonial future.[100] As a result, the playground might seem like an obvious site for teaching future generations the codes of normative citizenship. But beyond the provision of apparatus for strengthening exercise, there was rarely any instruction or guidance for children using such spaces. Whereas the US model required adults to organise the playground and to teach children how to play, this was generally not the approach adopted in Britain. While some advocates emphasised that a playground worker of the right background and temperament could help children to play properly, there is little evidence that adult play workers were a consistent feature of British public playgrounds at the turn of the century. As we will see in later chapters, it was only in the mid-twentieth century that such involvement became more commonplace. Instead, delivering Brabazon's geopolitical ambitions through the playground would instead rely on something like osmosis. The 'supervision of a judicious caretaker' would prevent 'tyranny and misconduct' but achieving broader objectives would not rely on the direct intervention of trained play workers – instead it would be achieved through self-directed exercise and a suitably green environment.[101]

As a short-term solution, Brabazon argued that vacant building plots could be turned into temporary spaces for children, equipped by the MPGA with simple gymnastic equipment until such time as the land was sold for development.[102] Despite raising the issue in parliament and organising well-attended local public meetings, the MPGA was largely unsuccessful in securing such short-term spaces for play. In contrast, efforts to create more permanent sites for children's recreation were considerably more successful, in part due to Brabazon's direct political influence. He was appointed as an Alderman of the LCC for a period of eight years in the 1890s and was also the first chair of its parks committee.[103] This helped to ensure not only a cooperative working relationship between the MPGA and local government officials but also the continued influence of Brabazon's vision for the playground. By 1892, the MPGA had made a direct financial contribution of over £14m (£27,991) towards the protection, acquisition or laying out of over fifty open spaces (in comparison, the figure for the Kyrle Society was eight).[104] Over the next decade the MPGA made significant further progress. By 1900 it had been involved in over one hundred sites and more than twenty included dedicated space for children. For example, the Association contributed over £1.4m (£3,000) towards the creation of Meath Gardens in Bethnal Green, which opened in 1894 and included two large playgrounds and a sandpit. Bartholomew Square (1895), near Old Street, had been 'asphalted for children especially' by the MPGA, who contributed £87,830 (£182) towards the cost.[105] In 1889 only two LCC parks, Myatt's Field and Finsbury Park, included specific facilities for children, but by 1915 thirty parks included children's gymnasiums.[106]

Beyond London, municipalities and philanthropic organisations were creating dedicated spaces for children in increasing numbers. In Manchester, the Prussia Street Recreation Ground opened in 1884 with see-saws and swings, and by the early twentieth century the provision of children's playgrounds in 'congested areas' such as Ancoats and Angel Meadow had become municipal policy.[107] Birmingham's Burbury Street Recreation Ground was covered with gravel by the borough surveyor in 1877 to make it suitable for use as a playground.[108] In 1914, the superintendent of parks in Edinburgh reported that fifteen children's gymnasiums had been provided in the city, while in Dublin four garden playgrounds equipped with apparatus had been established.[109]

This apparent progress conceals the difficulties sometimes associated with playground creation. St Paul's Churchyard and Playground on Rotherhithe Street in London was asphalted, equipped with gymnastic apparatus and set aside for children. It was opened by the MPGA in 1885, closed in 1888 only to be reopened in 1890 by the LCC. At St Leonard's Churchyard in Shoreditch 'the conduct of children was very bad', while in Dublin and London authorities established comprehensive regulations in an attempt to govern the use of playground spaces.[110] Even if they could be regulated, such play spaces did not always live up to campaigners' expectations, particularly those who were most interested in the benefits of nature and aesthetics. MPGA member Isabella Holmes felt that the new play space at Spa Fields provided a pale imitation of the pastoral version of nature that was most needed in the city. Even after the considerable money and effort, the playground was 'about as different from an ordinary village green, where country boys and girls romp and shout, as two things with the same purpose can well be'.[111] For Holmes, the green grass, cackling geese and picturesque cottages that surrounded the village green were replaced in the urban playground by gritty gravel and stray cats, encircled by dirty and monotonous housing.

Holmes's impression of the reworked Spa Fields demonstrates how bucolic landscape ideals were often more difficult to implement in the smaller spaces created by the MPGA. Reformers may have clung on to the potential of nature in the city, but those creating playgrounds were increasingly focused on the provision of natural features at a more manageable scale. Rather than expansive lawns, the MPGA promoted trees and shrubs as a pragmatic response, and it provided lists of specimens suitable for smoke-laden urban environments. Park authorities in Manchester went further in maximising room for play and 'very little attempt' was made to plant greenery in playgrounds located in the poor central districts of the city.[112] This tension between space for play and space for nature would be evident in playground discourse throughout the twentieth century. Despite the apparent disconnect between the imagined ideal and the reality on the ground, playgrounds were invariably popular and busy. After school hours and at the weekends, Spa Fields was full of children 'running about all over the open part of the ground', while further east the children's gymnasium in Victoria Park was described as popular, greatly used and often very crowded.[113]

In summarising the mid-nineteenth-century experience, we find that dedicated spaces for children's recreation had appeared in a small number of places intended to improve the lives of the urban poor. In Manchester and Salford, some of the earliest public parks included specific amenities for children, but these facilities were hardly central to the design or function of such spaces. In general, children were not a central constituent of the mid-century public park. Grounds for children's recreation were occasionally a feature of attempts to improve working-class housing, including Freshwater Place in London, but Charles Dickens's high-profile attempt to create more dedicated spaces for poor city children met with little success. Furthermore, children's place in the playground was far from certain as the term was still widely used to represent a range of spaces intended for adult recreation. The mid-century experience highlights the provisional nature of state, voluntary and philanthropic efforts to shape children's play through adaptations to the urban environments.

By the 1880s, changing attitudes to childhood, particularly the impact of compulsory education in conversely shaping time for recreation, created a wider social milieu that was more receptive to the need for dedicated public spaces for children. At the same time, wider anxiety about the social and political consequences of poverty and a burgeoning interest among philanthropic reformers in the problems of the urban environment focused attention on the pragmatic possibilities of the playground as part of wider efforts to reform the city and its inhabitants. Within this context, the Metropolitan Public Gardens Association combined urban anxieties and concern for the future of the empire with a commitment to the naturalistic public park and a belief in the positive potential of healthy and strong working-class children. Achieving this promise required a modification to earlier park ideals, particularly in relation to the location and size of green spaces and a greater emphasis on energetic, gymnastic exercise as the best route to health. Brabazon's role in both the MPGA and London County Council ensured that this vision for the playground featured in the thinking and action of both state and philanthropic actors in the capital. At the same time, delivering these objectives on the ground was rarely straightforward. The public spaces created by the MPGA often required significant work to provide a suitably level and hardwearing surface and the installation of gymnastic equipment to promote regenerative energetic exercise. Although sometimes supervised, adult involvement was primarily limited to caretaking, in contrast to US playgrounds and despite considerable transatlantic exchange.

But while the provision of dedicated recreational spaces for children gained credibility, particularly among increasingly confident municipal authorities, the influence, financial resources and effectiveness of the MPGA gradually declined. In part this was because Brabazon withdrew from public activities during the interwar period, but it might also be a product of the Association's apparent success in putting green space provision on the municipal map. Indeed, the principle of the children's playground was widely adopted by progressive local authorities after the First World War. But, as we shall see in the next chapter, while the 'idea' of the playground became more firmly embedded in the minds of park superintendents and urban reformers, its particular form in public parks and on housing estates was far from settled.

Notes

1. David A. Reeder, 'The Social Construction of Green Space in London Prior to the Second World War', in *The European City and Green Space: London, Stockholm, Helsinki and St. Petersburg, 1850–2000*, ed. Peter Clark (Aldershot: Ashgate, 2006), pp. 41–67; Matti O. Hannikainen, *The Greening of London 1920–2000* (Farnham: Ashgate, 2016).

2. Felix Driver, 'Moral Geographies: Social Science and the Urban Environment in Mid-Nineteenth Century England', *Transactions of the Institute of British Geographers*, 13 (1988), 275–87; Peter Thorsheim, 'Green Space and Class in Imperial London', in *The Nature of Cities*, ed. Andrew C. Isenberg (Rochester, NY: University of Rochester Press, 2006), pp. 24–37; Karen R. Jones, '"The Lungs of the City": Green Space, Public Health and Bodily Metaphor in the Landscape of Urban Park History', *Environment and History*, 24 (2018), 39–58; Kevin Brehony, 'A "Socially Civilising Influence"? Play and the Urban "Degenerate"', *Paedagogica Historica*, 39 (2003), 87–106; Hugh Cunningham, *Children and Childhood in Western Society since 1500* (London: Routledge, 2005); Harry Hendrick, 'Constructions and Reconstructions of British Childhood: An Interpretative Survey, 1800 to the Present', in *Constructing and Reconstructing Childhood*, ed. Allison James and Alan Prout (London: Routledge, 2015), pp. 29–53.

3. Thomas Beames, *The Rookeries of London: Past, Present and Prospective* (London: Thomas Bosworth, 1850); Henry Mayhew, *London Labour and the London Poor* (London, 1851), I; George Godwin, *Town Swamps and Social Bridges* (London: Routledge, Warnes and Routledge, 1859).

4. *Highway Act*, 1835, http://www.legislation.gov.uk/ukpga/Will4/5-6/50/section/LXXII/enacted [accessed 10 January 2022]; Sabine Schulting, *Dirt in Victorian Literature and Culture: Writing Materiality* (London: Routledge, 2016).

5. Linda M. Austin, 'Children of Childhood: Nostalgia and the Romantic Legacy', *Studies in Romanticism*, 42.1 (2003), 75–98.

6. Friedrich Froebel, *The Education of Man*, trans. by W.N. Hailmann (Norderstedt: Vero Verlag, 2015).

7. Samuel Wilderspin, *A System of Education for the Young* (London: Hodson, 1840).

8. Jane Read, 'Gutter to Garden: Historical Discourses of Risk in Interventions in Working Class Children's Street Play', *Children and Society*, 25 (2011), 421–34.

9. Charles Dickens, 'London Pauper Children', *Household Words*, 1850, p. 551.

10. 'A Want of the Age', *Bell's Life in London*, 17 January 1858, p. 7.

11. 'Playground and Recreation Society', *Lloyd's Illustrated*, 22 May 1859, p. 7.

12. Henry Silver, 'Ragged Playgrounds', *Punch*, 1 May 1858, 181; 'The Advantage of Taking a Short Cut through a Court', *Punch*, 4 June 1859, 233; 'Playground and General Recreation Society', *Daily News*, 3 June 1858, p. 3.

13. *Recreation Grounds Act*, 1859 https://www.legislation.gov.uk/ukpga/Vict/22/27/contents/enacted [accessed 9 January 2023].

14. 'The Gardens of Lincoln's Inn Fields', *The Times*, 22 May 1860, p. 12.

15. Jones, 'The Lungs of the City', p. 56.

16. David Lambert, 'Rituals of Transgression in Public Parks in Britain, 1846 to the Present', in *Performance and Appropriation: Profane Rituals in Gardens and Landscapes*, ed. Michel Conan (Dumbarton Oaks: Harvard University Press, 2007), pp. 195–210 (p. 197).

17. Lambert, 'Rituals of Transgression', p. 204; Metropolitan Public Gardens Association, *Eighteenth Annual Report* (London, 1900), p. 53.

18. Thomas Chawner and James Pennethorne, 'Plan for Laying out the Proposed Eastern Park to Be Called Victoria Park', 1841, British Library, Maps.Crace XIX 43a; 'Albert Park', 1851, National Archives, WORK 32/424.

19. Lambert, 'Rituals of Transgression'.

20. Nan Dreher, 'The Virtuous and Verminous: Turn of the Century Moral Panics in London's Public Parks', *Albion*, 29 (1997), 246–67.

21. Tom Hulme, *After the Shock City: Urban Culture and the Making of Modern Citizenship* (Woodbridge: Boydell Press, 2019).

22. W.T. Marriott, *The Necessity of Open Spaces and Public Playgrounds in Large Towns* (Manchester: Manchester and Salford Sanitary Association, 1862), p. 4, Wellcome Collection.

23. Ordnance Survey, 'Manchester and Salford Town Plan Sheet 21', 1850.

24. Allan Ruff, *The Biography of Philips Park, Manchester 1846–1996*, School of Planning and Landscape Occasional Paper 56 (Manchester: University of Manchester Press, 2000).

25. 'Official Inspection of the Manchester Public Parks', *Manchester Guardian*, 19 August 1846, p. 6; *A Few Pages about Manchester* (Manchester: Love and Barton, 1850).

26. George Forrest, 'The Giant Stride, or Flying Steps, and Its Capabilities', *Every Boy's Magazine*, 1 March 1862, pp. 122–7 (p. 122).

27. Jessica Riskin, 'Machines in the Garden', *Republics of Letters*, 1.2 (2010), 16–43.

28. Katy Layton-Jones, 'Manufactured Landscapes: Victorian Public Parks and the Industrial Imagination', in *Gardens and Green Spaces in the West Midlands since 1700*, ed. Malcolm Dick and Elaine Mitchell (Hatfield: University of Hertfordshire Press, 2018), pp. 120–37.

29. Joshua Major, *The Theory and Practice of Landscape Gardening* (London: Longman, Brown, Green and Longmans, 1852), p. 196.

30. *A Few Pages about Manchester*, p. 31.

31. 'Salford Town Council Park', *Manchester Examiner and Times*, 13 November 1850, p. 6.

32. In presenting financial amounts, this book adopts the same approach as Roderick Floud's *Economic History of the English Garden*. Based on a methodology developed by an international team of economic historians, Floud uses an index of average earnings, given the significance of labour in landscape projects, to translate historical financial values into present-day amounts. Using the *Measuring Worth* online calculator and assuming that creating a playground similarly represents a labour-intensive construction project, here financial amounts are quoted at 2021 values followed by the historical amount in brackets, enabling a straightforward comparison across time. Roderick Floud, *An Economic History of the English Garden* (London: Allen Lane, 2019); Measuring Worth, 'Five Ways to Compute the Relative Value of a U.K. Pound Amount, 1270 to Present', 2023, www.measuringworth.com/ukcompare [accessed 6 June 2023].

33. David Baldwin, 'Major, Joshua (1786–1866), Landscape Gardener and Designer', *Oxford Dictionary of National Biography* (Oxford: Oxford University Press, 2004).

34. 'Opening of Stamford Park, Altrincham', *Manchester Times*, 30 October 1880, p. 7.

35. 'Opening of the Portsmouth Public Park', *Hampshire Telegraph*, 29 May 1878, p. 3.

36. Susan Lasdun, *The English Park: Royal, Private and Public* (London: Andre Deutsch, 1991).

37. 'Newspaper Chat', *Examiner*, 12 June 1825, p. 745; 'Advertisements', *La Belle Assemblée*, 1 July 1825.

38. 'Voelker's Gymnastics', *Examiner*, 11 December 1825, p. 787.

39. 'Prof. Voelker Pentonville Gymnasium', *The Lady's Magazine*, 31 July 1827, 392.

40. 'The London Gymnastic Institution', *The Times*, 5 January 1827, p. 4.

41. Peter Henry Clias, *An Elementary Course of Gymnastic Exercises* (London: Sherwood, Jones & Co., 1823), p. v.

42. Donald Walker, *British Manly Exercises* (London: Hurst, 1834).

43. 'Government Gymnasium', *Illustrated London News*, 29 April 1848, p. 283.

44. 'The Gymnasium, Primrose Hill', *The Penny Illustrated Paper*, 2 May 1863, p. 4.

45. 'The Regent's Park Gymnasium', *The Standard*, 21 August 1878, p. 2.

46. The British Academy, *Reframing Childhood Past and Present: Chronologies* (London: The British Academy, 2019).

47. Andrew Mearns, *The Bitter Cry of Outcast London: An Inquiry into the Condition of the Abject Poor (1883)* (London: Frank Cass, 1970), p. 16.

48. William Booth, *In Darkest England and the Way Out* (London: Salvation Army, 1890), pp. xxi and 295.

49. Robert Henderson, 'Things Made by Children', *Strand Magazine*, 1897, 752–62; Brehony.

50. Octavia Hill, *Letters to Fellow-Workers 1864 to 1911*, ed. Elinor Southwood Ouvry (London: Adelphi, 1933), p. 5.

51. Octavia Hill, *Homes of the London Poor* (London: Macmillan, 1883), pp. 90–92.

52. Benjamin Ward Richardson, *Hygeia: A City of Health* (London: Macmillan, 1876), p. 27.

53. W.H. Prior, 'The Queen in Victoria Park', *Illustrated London News*, 12 April 1873, p. 349.

54. London County Council, 'Parks and Open Spaces, Descriptions, By-Laws, Acts of Parliament, Regulations', 1894, p. 152, London Metropolitan Archives, LCC/CL/PK/01/104.

55. 'The Trifler', *The Sunday Times*, 11 April 1880, p. 7; Sophia Elizabeth De Morgan, *Memoir of Augustus De Morgan* (London: Longmans Green, 1882).

56. John Springhall, 'Brabazon, Reginald, Twelfth Earl of Meath (1841–1929), Politician and Philanthropist', *Oxford Dictionary of National Biography* (Oxford: Oxford University Press, 2004); P.W.J. Bartrip, 'Hart, Ernest Abraham (1835–1898), Medical Journalist', *Oxford Dictionary of National Biography* (Oxford: Oxford University Press, 2004).

57. Clare Hickman, 'To Brighten the Aspect of Our Streets and Increase the Health and Enjoyment of Our City', *Landscape and Urban Planning*, 118 (2013), 112–19.

58. Reginald Brabazon, *Social Arrows*, 2nd edn (London: Longmans, Green & Co., 1887), p. 13.

59. Jim English, 'Empire Day in Britain, 1904–1958', *The Historical Journal*, 49 (2006), 247–76.

60. William Coates, 'The Duty of the Medical Profession in the Prevention of National Deterioration', *British Medical Journal*, 1 (1909), 1045–50.

61. Reginald Bray, 'The Children of the Town', in *The Heart of Empire: Discussions of Problems of Modern City Life in England, with an Essay on Imperialism*, ed. Charles Masterman (London: Fisher Unwin, 1902), pp. 111–64 (p. 126).

62. Metropolitan Public Gardens Association, *Eighteenth Annual Report*, p. 31.

63. H.L. Malchow, 'Public Gardens and Social Action in Late Victorian London', *Victorian Studies*, 29 (1985), 97–124 (pp. 109–14).

64. Metropolitan Public Garden, Boulevard, and Playground Association, *Second Annual Report* (London, 1884), p. 9, US National Library of Medicine, archive.org, 101200449.

65. Ernest Hart, 'Graveyards as Recreation Grounds', *The Times*, 20 August 1885, p. 8.

66. Metropolitan Public Gardens Association, *Eighteenth Annual Report*, p. 32.

67. Metropolitan Public Gardens Association, *Twenty Second Annual Report*, 1904, The Guildhall Library, ST 317.

68. W.J. Gordon, 'The London County Council and the Recreation of the People', *The Leisure Hour*, 1894, pp. 112–15.

69. Marriott, *The Necessity of Open Spaces*, p. 7.

70. Walter Besant, 'The Social Wants of London: IV Gardens and Playgrounds', *The Pall Mall Gazette*, 18 March 1884, pp. 1–2 (p. 1).

71. Bray, 'The Children of the Town', pp. 116–18.

72. Brabazon, *Social Arrows*, p. 54.

73. Earl of Meath, 'Public Playgrounds for Children', *The Nineteenth Century*, 34 (1893), 267–71.

74. Archibald Maclaren, *A Military System of Gymnastic Exercises* (London: HMSO, 1868).

75. Mathias Roth, *Gymnastic Exercises without Apparatus According to Ling's System*, 7th edn (London: A.N. Myers & Co., 1887).

76. Gertrud Pfister, 'Cultural Confrontations: German Turnen, Swedish Gymnastics and English Sport – European Diversity in Physical Activities from a Historical Perspective', *Culture, Sport, Society*, 6 (2003), 61–91.

77. Metropolitan Public Garden, Boulevard, and Playground Association, *Second Annual Report*, p. 23.

78. Elizabeth Crawford, 'Wilkinson, Fanny Rollo (1855–1951), Landscape Gardener', *Oxford Dictionary of National Biography* (Oxford: Oxford University Press, 2008).

79. Madeline Agar, *Garden Design in Theory and Practice*, 2nd edn (London: Sidgwick & Jackson, 1913), p. 200.

80. Earl of Meath, *House of Lords Debate, 9 June 1890, Vol.345, Col.264–267* (Hansard, 1890).

81. See, for example, Thomas Firmin to Poor Law Board, Letter, 22 August 1854, National Archives, MH 12/11000/243, fo.449–50; Charles Hart to Poor Law Commission, Letter, 14 November 1845, National Archives, MH 12/5967/153, fo.300.

82. Warwick Wroth, *The London Pleasure Gardens of the Eighteenth Century* (London: Macmillan, 1896).

83. Peter Thorsheim, 'The Corpse in the Garden: Burial, Health, and the Environment in Nineteenth-Century London', *Environmental History*, 16 (2011), 38–68.

84. John James Sexby, *The Municipal Parks, Gardens, and Open Spaces of London: Their History and Associations* (London: Elliot Stock, 1898).

85. London County Council, 'Ceremony of Opening Little Dorrit's Playground, Southwark', 1902, London Metropolitan Archives, LCC/CL/CER/03.

86. W.W. Pettigrew, 'Park Superintendent's Report Book', trans. by Anne Bell and Andrew Bell, 1908, Cardiff Council Parks Service, uncatalogued.

87. 'Children's Playground in South London', *Illustrated London News*, 10 May 1884, p. 443; Metropolitan Public Garden, Boulevard, and Playground Association, *Second Annual Report*, p. 19.

88. An almost identical approach was followed by the Metropolitan Free Drinking Fountain Association (1859) in its efforts to provide public drinking fountains (see Metropolitan Free Drinking Fountain Association, *Annual Report*, 1865, London Metropolitan Archives, ACC/3168/017).

89. Metropolitan Public Gardens Association, *Annual Report*, 1932, London Metropolitan Archive, CLC/011/MS22290.

90. London County Council, 'Return of the Names and Wages of and Work Performed by All Persons Employed in the Council's Parks and Also of the Respective Areas Devoted to Gardens, Lawns, Fields and Playgrounds and the Extent of Conservatories in Such Parks', 1889, London Metropolitan Archives, LCC/CL/PK/01/104.

91. London County Council, 'Report of the Parks and Open Spaces Committee, 16 May', 1893, p. 5, London Metropolitan Archives, LCC/CL/PK/01/104.

92. Frederick Law Olmsted, *Public Parks* (Brookline, MA, 1902), p. 107; Earl of Meath, *Public Parks of America*, Report to the Parks and Open Space Committee, London County Council, 1890, London Metropolitan Archives, LCC/PUB/02/01/066; John Lucas, 'A Centennial Retrospective: The 1889 Boston Conference on Physical Training', *Journal of Physical Education, Recreation & Dance*, 60.9 (1989), 30–33.

93. Henry S. Curtis, *Education Through Play* (New York: Macmillan, 1915), p. 85.

94. Rowland Hayes to Paul C. Wilson, 'Papers from New York Committee on Recreation to the New York City Mayor's Office', 29 October 1917, National Archives, RECO 1/694.

95. Everett B. Mero, *American Playgrounds: Their Construction, Equipment, Maintenance and Utility* (Boston: School of Education, Harvard University, 1908), p. 23.

96. Frederick Law Olmsted, 'Boston: Charleston Playground: General Plan', 1891, Artstor/University of California, San Diego.

97. 'An Organized Playground', *The Times*, 8 July 1909, p. 9.

98. Walter Wood, *Children's Play and Its Place in Education* (London: Kegan Paul, Trench, Trübner & Co., 1913).

99. B. Holland, 'London Playgrounds', *Macmillan's Magazine* (London, 1882), XLVI edition, pp. 321–4.

100. Carole O'Reilly, 'From "The People" to "The Citizen": The Emergence of the Edwardian Municipal Park in Manchester, 1902–1912', *Urban History*, 40 (2013), 136–55.

101. Metropolitan Public Gardens Association, *Eighteenth Annual Report*, p. 31.

102. Brabazon, *Social Arrows*, p. 40.

103. 'London County Council', *The Times*, 6 February 1889, p. 11; 'London County Council', *The Times*, 30 March 1898, p. 15; 'Opening of Clissold Park', *The Standard*, 25 July 1889, p. 3; 'Lord Meath's Memories', *The Times Literary Supplement*, 10 May 1923, p. 317.

104. London County Council, 'County of London Parks, Open Spaces and Commons', 1892, London Metropolitan Archives, LCC/CL/PK/01/104.

105. Metropolitan Public Gardens Association, *Eighteenth Annual Report*, pp. 39 and 57, see appendices for a full list of sites and the MPGA's contribution.

106. London County Council, 'Return of the Names and Wages'; London County Council, 'Parks and Open Spaces: Regulations Relating to Games, Together with Particulars of the Facilities Afforded for General Recreation', 1915, LSE Library, 421 (129D).

107. Hazel Conway, *People's Parks: The Design and Development of Victorian Parks in Britain* (Cambridge: Cambridge University Press, 1991); W.W. Pettigrew, *Handbook of Manchester Parks and Recreation Grounds* (Manchester: Manchester City Council, 1929), p. 8.

108. 'Opening of Burbury Street Recreation Ground', *Birmingham Daily Post*, 3 December 1877, p. 8.

109. John W. McHattie, *Report on Public Parks, Gardens and Open Spaces* (Edinburgh: City of Edinburgh, 1914), RHS Lindley, 999 4C EDI; Vanessa Rutherford, 'Muscles and Morals: Children's Playground Culture in Ireland, 1836–1918', in *Leisure and the Irish in the Nineteenth Century*, ed. Leeann Lane and William Murphy (Liverpool: Liverpool University Press, 2016), pp. 61–79.

110. Metropolitan Public Gardens Association, *Eighteenth Annual Report*, p. 45; Rutherford, 'Muscles and Morals'; London County Council, 'Parks and Open Spaces, Descriptions, By-Laws, Acts of Parliament, Regulations'.

111. Isabella M. Holmes, *The London Burial Grounds: Notes on Their History from the Earliest Times to the Present Day* (London: T.F. Unwin, 1896), p. 277.

112. Pettigrew, *Handbook of Manchester Parks*, p. 22.

113. Holmes, *The London Burial Grounds*, p. 276; London County Council, 'Report as to the Condition of Victoria Park', 1893, London Metropolitan Archives, LCC/CL/PK/01/104.

Chapter 2

Competing playground visions: 'a distinctly civilizing influence that gives much health and happiness'

In the first decades of the twentieth century, the children's playground became more firmly embedded in visions for a better urban environment. Similarly, the provision of dedicated public spaces for play formed part of wider calls for more comprehensive action by the state to address social problems. The history of the playground provides a complex account of evolving social policy and urban interventions at this time, one characterised by a mix of philanthropic and state action and claims to expertise by commercial and campaigning organisations. These diverse claims resulted in competing visions for the spaces where children were supposed to play and notable public debate about what constituted a 'properly equipped playground'. In exploring the changing fortunes of these visions in the early decades of the twentieth century, this chapter considers several important themes. On the one hand, it charts an ongoing belief in the health benefits of outdoor exercise and education, despite changing scientific ideas about the spread of disease. Although miasma had become less credible as a vector of ill health, there remained a widely held belief in the benefits of exercise and education in the healthy open air. As experimental schools were built without walls to ensure pupils' full exposure to the elements, the outdoor and energetic public playground was imbued with a renewed sense of healthiness. On the other hand, the playground featured in evolving ideas of social welfare, philanthropic advocacy and commercial opportunism, which came together in this period to shape the urban environment. As ideas about play space design crystallised, manufacturers of playground apparatus worked with municipal authorities to deliver playgrounds on the ground in greater numbers, pointing to the unusual place of the playground within wider historical analysis of the democratisation and commercialisation of leisure in this period. A case study of Charles Wicksteed, his manufacturing company and the park he created in Kettering, Northamptonshire, highlights the diverse processes involved and the ongoing significance of philanthropy, voluntary action and commerce in shaping public spaces for play. As such, the chapter plots a key moment in the history of the playground, charting changing social values in relation to children's play and the considerable influence of playground equipment suppliers and amusement park rides on notions of the ideal playground form.

'Properly equipped playgrounds' in the early twentieth century

At the turn of the century, advocates of the transatlantic playground felt that providing apparatus for entertainment was the least important element in a playground. In the USA, Henry Curtis argued that 'the thing of first importance is organization; next in importance is equipment for games; next comes provision for athletics; and last such apparatus as swings and slide'.[1] In contrast, British commentators increasingly imagined the playground as a space where equipment was a central feature. The pioneering garden historian Alicia Amherst argued in 1907 that a fully equipped public park needed to include not just high-quality horticulture but also swings and other gymnastic equipment for children.[2] For the author and journalist Annesley Kenealy, also writing in 1907, the 'pitilessly meagre surroundings' of the gravelled children's playground in St James's Park did little to save children from the 'unwholesome sights and sounds of a sordid, huckstering, fetid slum street'. Instead, a properly equipped playground was needed in every park, including amenities such as seating, a water fountain, sandpit, low swings, see-saws, horizontal bars and a giant stride.[3] In 1909, an anonymous letter writer to *The Times* concurred, suggesting that the nearby Kensington Gardens needed at least another dozen swings and a shallow pond for paddling, plus an end to Sunday closing of playgrounds more generally.[4]

This emphasis on the park-based playground was a result of the continuing anxiety about the health of working-class children, the perceived problems of the urban environment and an enduring faith in the curative potential of urban green space. Medical ideas about the cause of disease had been shifting since the late nineteenth century, moving away from a belief in the role of miasma or bad air as the main vector for ill health, towards germ theory where specific organisms caused disease. At the same time, the power of the public park to provide metaphorical lungs for the city and fresh air for its inhabitants remained powerful. These socio-medical beliefs were given material expression in the development of open-air schools, often in or near parks, and the restorative potential of open-air treatments for illnesses including rickets and tuberculosis. In London, experiments in open-air education had started in 1907 adjacent to Horniman Park, while in Nottingham, open-air schools were built in several public parks where sunlight and fresh air would improve the health of the children attending.[5] At Cropwood School in the West Midlands, open-air classrooms, a sleep garden, outdoor swimming pool and playing lawn all facilitated the combination of clean air, playful exercise and enforced rest needed to treat illness and restore pupils' physical and moral health.[6] Beyond this belief in the power of fresh air to treat specific diseases, exposure to sunlight and outdoor recreation were also positioned as healthy activities more generally. For instance, park-based lidos provided the wider population with opportunities for outdoor swimming and exposure to the sun, and they were built in increasing numbers by municipal authorities.[7]

Within this wider social and cultural context, a belief in the benefits of open-air physical exercise remained a powerful and widespread justification for playground provision. For the youth worker Charles Russell, speaking at a meeting of the Manchester and Salford Playing Fields Society, dedicated space for children was vital to 'check the degeneration which any overcrowded area in the kingdom could show'.[8] The Liberal politician and author Charles Masterman may have disagreed with the imperial politics of Brabazon and the MPGA, but he also emphasised the problematic association between the unhealthy environment and urban childhood. The twice-breathed air and disconnect from nature resulted in 'the production of a characteristic *physical* type of town dweller: stunted, narrow chested, easily wearied; yet voluble, excitable, with little ballast, stamina or endurance'.[9] An appropriately sited playground could help to tackle many of these physical and moral issues.

In practice, inserting dedicated open-air spaces for children into the urban environment was far from straightforward and advocates needed to negotiate a route through competing expectations of public space that were shaped by notions of age, class and gender. The way that park administrators responded to these calls for playground improvements provides an insight into the factors that informed the way that spaces changed on the ground. In 1909 and in response to newspaper articles, Kensington Gardens' administrators attempted to provide facilities and shape regulations that balanced the needs and expectations of a wide variety of groups.

In establishing Sunday opening times, officials needed to find a balance between public pressure and the views of influential religious groups. In assessing the need for more paddling pools, they set out to find a balance between providing an appropriately equipped place for children to play, and the risk that they would be 'continually receiving complaints from irate parents' when children fell into the water. The provision of a new sandpit would put a strain on finances, but parks staff demonstrated their willingness to use the media to their advantage, suggesting that an anonymous letter to newspapers might solicit a private donor to pay for new facilities. They had much less trouble in fixing gender- and age-related boundaries; park keepers were warned to 'prevent grown women using the present swings'.[10]

But just as calls were being made to improve playgrounds by adding appropriate equipment and facilities, the principle of providing space for outdoor, energetic exercise was being questioned. By 1909, the Primrose Hill gymnasium (visited in the previous chapter) was seen as improperly equipped and acted as a focus for debates about what was considered to be legitimate use of equipped public space. The gymnasium caused practical and moral problems for park managers, who had to navigate a path between the differing expectations of wealthy neighbours and the gymnasts that used it. Its drinking fountain was 'a source of constant annoyance' as children splashed passersby. On another occasion, the entire gymnasium was closed at the urgent request of the police, as a result of unseemly language, rowdyism, and the misuse of the space as 'a training ground for prostitutes'. While a subsequent petition called for the gymnasium to be re-opened, petitioners also complained that a proper gymnasium should be indoors and accompanied by changing facilities and appropriate instruction.[11]

Such anxieties over the appropriate use of public space, and wider concern about the health of the population, were exacerbated by the First World War and its impact on the home front. Large numbers of working-class conscripts were exempted from military service as a result of physical unfitness, with over a million rejected on medical grounds in the last year of the war.[12] Furthermore, absent fathers, children working in munitions factories far from home and even the cinema were blamed for a perceived rise in juvenile delinquency.[13] The perceived contribution of such misbehaviour to a shell shortage in 1915 prompted the government to establish Juvenile Organisations Committees who would channel the work of existing philanthropic and voluntary organisations towards the wartime military objectives of the state. Although primarily established to structure the leisure time of adolescents and young adults, the committees symbolised the beginnings of a shift towards promoting welfare, rather than criminalising young people.[14] For Lord Lytton, chair of the State Children's Association, a system of 'reclamation through friendship' rather than resorting to the courts was the solution to the problems of youth.[15]

The provision of playgrounds for younger children seems to have proceeded in this vein too, as play provision became associated with the broader welfare of children and their families. During the war, the philanthropic Carnegie UK Trust appointed the noted physician and medical officer Janet Campbell to undertake a comprehensive investigation into the health and wellbeing of mothers and young children.[16] Campbell's influential report, published in 1917, examined in detail the provision of midwifery services, nurseries and play schemes, as well as the playgrounds' role in improving the welfare of children and their parents. Echoing Masterman's view from fifteen years earlier, Campbell argued that the lack of suitable play opportunities contributed to defective child development. In contrast, the provision of appropriate play spaces and activities would ensure the proper physical and mental development of children and prevent juvenile delinquency by providing an alternative to the street for children's recreation.[17] A similar study in Scotland by the noted public health administrator Leslie Mackenzie found that time on the playground seemed to have direct medical benefits too. For one medical observer who contributed to the study, playgrounds helped to tackle runny noses and improved children's nutrition, in addition to the more commonly ascribed physical benefits of open-air exercise and toning muscles. The Scottish study in particular noted the importance of providing both better working-class housing and better places to play, concluding that 'the toddler's playground is fundamentally essential to the health of the children that occupy the crowded quarters of every city. The open-air playground is the counteractive to the poisonous house.'[18] Dedicated outdoor space would help to create healthier children and a more salubrious urban environment.

The notion that leisure time should be spent constructively, as park advocates had imagined in the nineteenth century, had not been superseded entirely. The 1919 National Conference on the Leisure of the People demonstrated increasing concern with the 'possibilities for good or evil' associated with increasing time for rest among the working-class: 'if rightly used it will be in these hours the growing boy or girl will receive that wider education which is going to build character, make him an intelligent workman and a useful citizen'.[19] At the same time, technology, commerce and democratisation also meant that leisure could be idly rather than constructively spent.[20] However, using these processes as a way of understanding the changing nature of children's playgrounds is not straightforward. Leisure was largely constructed in opposition to work, and as social constructions of childhood no longer included work this meant that ideas about new forms of leisure were invariably adult-centric. Democratisation implies that people had an increased say in how and where they participated in leisure activities, but neither the principle of the playground nor the way that it was designed meant that this was the case for children. There is no evidence to suggest that children were given a say in where play spaces were located or how they were designed, nor whether they were the spaces where children preferred to play. While it is difficult to see playgrounds as spaces of democratisation for children, they were nonetheless affected by processes of commercialisation, although in different ways to other aspects of postwar leisure provision.

The historian Peter Borsay has concluded that there have been commercial aspects to leisure since early modern times, but that more recent commercialisation has been associated with increased demand driven by rising disposable incomes.[21] This assumes that individual participants are directly purchasing leisure opportunities. In the case of a seemingly non-commercial space such as a playground, demand has been driven by social and cultural factors, while supply has been driven by commercial ones. The economic wealth of potential playground users did not create demand for playgrounds; from a commercialisation of leisure perspective, playgrounds are unusual in that they have been free to visitors at the point of use. Instead, demand for playgrounds was the result of evolving social ideas about childhood and public space and the associated targeting of philanthropic and municipal funding. Urban municipalities, for example, increased spending on parks and open spaces from £11,830 (£93) per thousand of population in 1920 to £25,270 (£131) by 1929.[22] Commercialisation in this case was mediated through philanthropists, park superintendents and municipal administrators who purchased and created leisure spaces for children. As in other aspects of leisure provision, the creation of play spaces and the supply of appropriate apparatus had been shaped by technology, entrepreneurialism and professionalisation. For playgrounds in particular, a diverse range of manufacturing companies, including gymnastic outfitters, fencing companies and engineers, were entrepreneurial in applying and adapting their existing technologies and production lines to meet the requirements of an increasingly organised and specialised parks profession.

The historian Thomas Richards has argued that the increased availability of manufactured goods in the nineteenth century, along with new forms of marketing, fostered a national culture of consumerism.[23] While this saw a significant expansion in the use of retail catalogues to sell goods to the public, including plants and other gardening equipment, it was commercial catalogues that shaped the creation of playground spaces in the early twentieth century. Historian Claire Jones has shown how commercial catalogues helped to shape both knowledge and practice among medical professionals, mediating between the seemingly incompatible spheres of commerce and professionalism.[24] Although play equipment catalogues, and their professional and commercial context, were very different to medical ones, manufacturers and their marketing materials nonetheless made a significant contribution to the evolution of the playground ideal and shaped professional approaches to the design and creation of dedicated spaces for children's leisure.

In 1923, a member of the public, Hubert Seligman, took the unusual step of writing to the royal parks, offering to purchase see-saws for installation in St James's Park.[25] Seligman was from an Anglo-American merchant banking family with a commitment to philanthropic activity and lived close to Kensington Gardens and Hyde Park.[26] In response, park administrators sought prices for see-saws from a number of manufacturers and received quotes and catalogues in reply. Seligman bought two see-saws, at a cost of £5,233 (£24 5s), and would go on to regularly offer specific items of play equipment for different London parks over the next ten years. This was an unusual example and playground provision was rarely driven by direct requests such as this, but the story does provide a useful insight into the processes, people and objects involved in the production of playground spaces in the early twentieth century.

Approaching commercial suppliers suggests that park superintendents could not call on inhouse skills or experiences to design and build their own gymnastic or playground apparatus. While the creation of new horticultural schemes each year meant that a plant nursery and the associated staff were a worthwhile investment, the infrequent need for new playground equipment meant there was little value in investing in inhouse manufacturing technology or infrastructure. With their professional background in horticulture, landscape design or public administration, park superintendents perhaps felt they lacked an understanding of the needs of children, let alone an awareness of contemporary educational or theoretical thinking about childhood and play.

In response, manufacturing companies attempted to present their catalogues as informative and educational documents, a source of expertise on children's playgrounds. Spencer, Heath and George of London promoted the fact that they could send an 'expert representative, free of charge' to view a potential play space and would prepare a specific playground scheme for customers. The implication was that they knew what a playground should be, what children needed and also that their version of the playground was the norm. By using the adjective *regulation* when asserting that they supplied 'regulation playground outfits', they were presumably attempting to show that their products met with long-established ideas about the playground (Figure 2.1). Their catalogue included images of their apparatus installed in Newtongrange Public Park in Midlothian and listed the LCC parks where their products had been installed. In a similar way, the catalogue supplied by Bayliss, Jones and Bayliss of Wolverhampton included photos of their equipment in a City of Birmingham open space, as well as a list of apparatus that would create an ideal playground (Figure 2.2). By including images of their equipment in existing open spaces, both companies were attempting to legitimise and extend their particular version of the children's playground.[27]

The production of playground equipment seems to have been a by-product for these companies. Manufacturers drew upon their existing technological knowledge, adapting existing products and production lines to take advantage of the new business opportunities. Their playground creations seem to have been reconfigured versions of their other products, making use of familiar materials and manufacturing processes. Spencer, Heath and George described themselves as gymnastic outfitters and manufacturers of calisthenic gear, gymnasium buildings and boxing rings. While creating versions of their products for outdoor use was likely to be a logical and relatively straightforward step, it was also something that to an extent they made up as they went along; responding to a letter from the Regent's Park superintendent in 1924, they were unable to provide a drawing or photo of their see-saw and instead sent a roughly drawn, free-hand sketch with somewhat clumsy annotations.[28]

In a later, more professional-looking catalogue, Spencer, Heath and George also emphasised the technological superiority of their products – their plank swing included 'self-aligning roller bearing fitments' which meant they felt able to claim it was 'mechanically perfect'.[29] Bayliss, Jones and Bayliss's catalogue, *Gymnasia for Parks and Recreation Grounds*, clearly showed where the focus on their manufacturing business lay – nearly half of their catalogue is dedicated to the variety of fencing, guard rail and entrance gates they produced, while their sketch of the ideal playground includes significant lengths of fencing on all sides.[30] Even individual items of equipment appear to make use of the materials and forms of fencing components.

When manufacturers claimed to have 'expert representatives' it seems most likely that they were experts in the products that the companies sold, rather than anything else. Although this expertise was unlikely to be grounded in the emerging professional and academic ideas about

Figure 2.1: Regulation playground outfits, Spencer, Heath and George Ltd, no date, National Archives, WORKS/16/1705.

Figure 2.2: Gymnasia for parks, Bayliss, Jones and Bayliss Ltd, 1912, National Archives, WORKS/16/1705.

child development of the time, it was able to deliver a particular version of the playground with its roots in contemporary attitudes towards age, gender and exercise. Fencing manufacturers could provide products to enclose dedicated spaces for children to play. Gymnastic outfitters could provide apparatus that could direct children to take part in particular forms of physical exercise that would have beneficial consequences for both individuals and society.

As well as building on manufacturers' existing technological knowledge, the catalogues also emphasised the benefits that their products could offer to their customers (although rarely the benefits they might offer to children). As a result, they provide an insight into manufacturers' perceptions of park superintendents' concerns, values and assumptions, as well as wider social values about children and their use of public space. As well as emphasising the technological innovation of products, play equipment catalogues consistently played upon three key narratives – firstly, the risk of deliberate damage by children to the playground; secondly, the need for playgrounds to be safe for the children using them; and thirdly, the segregation of play spaces by age and gender.

The perceived threat of hooliganism and the associated nuisance, danger and moral consequences reflected an anxious mood in late Victorian and Edwardian Britain.[31] Equipment manufacturers were able to play on this anxiety when advertising their products. Catalogues presented a sanitised and choreographed version of children's play, where text and images emphasised the robustness of equipment in the face of potential damage, as well as the beneficial effects of a properly equipped playground in maintaining order and respectable behaviour. The strength and durability of apparatus was emphasised, and in some cases explicitly guaranteed as hooligan-proof, while tree guards, strong seats and unclimbable railings would limit the opportunities for children to damage other features. The photos in Bayliss, Jones and Bayliss's catalogue show clean, respectably dressed children, posing on stationary equipment or awaiting their turn in an orderly queue. A policeman is present in the background of all wider photos of the playground, providing added reassurance to potential customers that a Bayliss-equipped playground would be an orderly place, but also hinting at the disorder that was possible.[32]

A second, interrelated rhetoric employed by manufacturers emphasised safety, primarily in relation to the way that children used playground equipment. For example, the term see-saw seems to have been applied to a physical structure, rather than just the associated up-and-down motion, as early as the 1820s.[33] By the late nineteenth century, there was increasing anxiety that the sudden bump of a see-saw onto the ground could hurt not only children's feet but also damage their spines, leading several commentators to describe them as one of the most dangerous and accident-prone items on the playground.[34] This focus on the risk of spinal injury in particular echoed evolving ideas about the spine as a conduit for physical and mental health. In particular, the increasing number of people suffering from a condition known as railway spine, which saw some passengers involved in railway accidents suffering no physical injuries but subsequently developing debilitating nervous shock, and the high-profile coverage of associated court cases perhaps made spinal injuries particularly worrisome.[35] Whether the manufacturers' emphasis on safety was born of anxiety for children's physical and mental health or as a result of the financial compensation paid by the railway companies to injured passengers is not clear, but they did stress the safety of their equipment nonetheless. Each company emphasised the adaptations to their products that would help to ensure children's wellbeing, from air cylinders that dampened see-saw impacts to safety tails on slides that prevented a sudden fall to the ground at the bottom. Despite manufacturers' claims, park superintendents had to purchase and install equipment in playgrounds before they could assess for themselves its safety in use. For example, royal parks staff annotated Spencer, Heath and George's promotional drawings, noting that both their version of the see-saw and the giant stride were still 'found to be dangerous in practice'.[36]

It was not only equipment that posed a risk to children in the playground; the inappropriate behaviour of adults was an issue that manufacturers also sought to address. In 1913, the Metropolitan Radical Federation highlighted the 'frequent indecent offences towards children' in the parks of London and called for more park keepers who could detect and prevent such offences.[37] A year later, the LCC education committee also emphasised 'the evils which appear to arise owing to the lack of adequate supervision'.[38] Royal parks administrators empathised with the malevolence of the offences, but felt that the relatively small number of reported incidents – on average nine per year across all the royal parks in London – meant that they could not justify increasing the number of plain clothes staff on duty to detect such offences, and in any case doubted the effectiveness of such an action.[39] Commercial playground manufacturers attempted to provide a solution to the problem by supplying gates and fencing that could exclude undesirable adults from the playground, although the efficacy of such an approach was likely to be questionable.

In addition to attempts to separate children and adults, there were also efforts to segregate girls and boys when using the playground. Elizabeth Gagen has shown how early twentieth-century play spaces helped to reproduce conservative gender politics in the USA, a process that can be seen in the actions of both park authorities and equipment manufacturers in Britain too.[40] In 1904, the LCC provided separate gymnasiums for girls and boys in thirteen of the open spaces it managed, including Spa Fields and Meath Gardens, while Victoria Park, Battersea Park and a further eight green spaces included gymnasiums for exclusive use by girls.[41] This physical segregation of play spaces was something that manufacturers were easily able to support. For example, Bayliss, Jones and Bayliss could provide fencing to divide playgrounds for boys and girls, while a number of catalogues also included gender-specific products. However, closer inspection shows marginal differences between such items. Bayliss's swings for girls were thirteen feet high, cost £13,120 (£31 7s 6d), included seats and had a sign on top which said *For Girls Only*. Swings for boys were the same height, cost the same and also included seats. The main differences were that swings for boys could also be fitted with trapeze bars and rings and the sign on top which said *For Boys Only*. Similarly, Bayliss's list of equipment for the ideal playground for girls was remarkably similar to that for boys, with the exception that a boys' playground needed one of everything from the catalogue, at a cost of £45,540 (£108 17s 6d), while in an ideal girls' playground the vaulting horse was replaced by three see-saws. Just in case the swing signage or loitering police officer proved ineffective, Bayliss were also able to supply 5-foot-high 'wrought iron unclimbable railings' for £139 (6s 8d) per yard to keep children apart.[42] Just surrounding the swings, let alone the whole playground space, would add half again on top of the cost of the equipment; the perceived need to segregate girls and boys at play had direct economic, as well as political and personal consequences.

While early playground equipment manufacturers may have echoed wider social attitudes relating to the use of public space, they also seem to have been inspired, in part at least, by other types of amenity landscape which emphasised enjoyment and delight and incorporated technological innovation. As early as 1835, the noted garden designer John Claudius Loudon described in his *Encyclopaedia of Gardening* a number of European aristocratic estates that included temporary or permanent swings and roundabouts.[43] The great exhibitions and world fairs of the nineteenth century, while ostensibly educational, were often more commonly experienced by visitors as spaces of entertainment.[44] There had also long been an association between green space and commercial leisure provision, most notably in the eighteenth-century pleasure gardens such as Vauxhall.[45] However, the spectacular performances, nocturnal illuminations, notorious immorality, entry fees and most significantly the prohibition of children suggest significant differences between such spaces and the emerging children's playground.[46]

In contrast, there were more direct connections between the Edwardian amusement park landscape and the form of dedicated spaces for children. Alongside circus acts and novelties, swings were a regular feature of travelling and seasonal fairs and, as we saw in the previous chapter, privately operated fairground-style swings were located for a time in London's Victoria Park.[47] Merry-go-rounds also seem to have been a regular feature of travelling fairs, in Turkey from the seventeenth century and in Britain from the eighteenth century, and by the late nineteenth century they were often steam powered and elaborately decorated.[48] But it would be the amusement park, rather than the public park, where such temporary fixtures would become permanent installations, inspired by a transatlantic exchange of ideas.[49]

In the early twentieth century, P.G. Wodehouse associated the uncertain profits of the travelling showman with the motion of his fairground rides, so that any income lost on the swings might be made up on the roundabouts.[50] In doing so, Wodehouse also inadvertently connected the fluctuating fortunes of the playground ideal and circuitous themes in play space discourse with structures that would soon come to symbolise its presence in public space. The architectural historian Josephine

Kane has described how an assortment of rides at Blackpool South Shore became an American-style amusement park in 1903 and prompted a surge of schemes elsewhere, particularly in seaside resorts such as Margate, Southend and Great Yarmouth. The characteristic combination of noise, bright colours and frenetic movement, plus modern architecture and technologically produced sensations, made the early twentieth-century amusement park landscape a unique whirl of wonders.[51] And while such spaces legitimised childlike behaviour by adults, they also influenced the form of playground spaces too. The idealised bucolic city park was not transformed into the whirling landscape of the amusement park, but the technological innovation and sense of freedom that characterised the latter were influential in shaping a particular approach to play provision. As the next section shows, individual rides would be scaled down, simplified and introduced into the park playground, as would more accommodating attitudes towards the behaviour of both children and adults. One play equipment manufacturer in particular was at the forefront of these changing attitudes to public spaces for children and from the 1920s onwards promoted a particular vision for the children's playground that was at odds with earlier attempts at regulation and segregation.

Charles Wicksteed, philanthropy and commerce

Following Mr Seligman's offer to purchase a see-saw for St James's Park in 1923, one other manufacturer responded to the royal parks' request to supply information. Charles Wicksteed & Co., an engineering firm based in Kettering, sent a covering letter with information about the see-saw they could supply, along with a rudimentary catalogue. Like the other catalogues submitted, it included images of apparatus, prices and descriptions. In contrast, it also included a two-page preface, where Charles Wicksteed set out his personal vision for the ideal children's playground. He stated that his vision was based on his own experiences of creating and managing a public park and playground in Kettering. While this could be seen as – and perhaps was to some extent – a refined sales pitch, it is useful to understand his motives and the impact they had on the playground ideal because elements of this vision soon spread across the UK and around the world.

The commercialisation of leisure has been well documented, but much less has been done to explore the role of voluntary action in interwar leisure provision.[52] Although often associated primarily with nineteenth-century public parks, philanthropic involvement in the creation of civic green spaces remained important in the interwar period. For example, chocolatier Joseph Rowntree gifted a riverside park to the city of York in 1921 and as we have already seen Mr Seligman was donating individual playground structures throughout the 1920s.[53] The ideas and actions of Charles Wicksteed provide a noteworthy case study because they combine the processes of commercialisation and philanthropy within the public park. The creation of Wicksteed Park and its playground were characteristic of the voluntary action that sought to foster good citizenship through leisure, but at the same time the manufacture and sale of play equipment by Wicksteed & Co. was a commercial venture. Making sense of the philanthropic motives and political assumptions that underpinned Wicksteed's actions, as well as the role of his manufacturing company, helps to shed light on the processes involved in shaping popular and professional notions of what constituted an appropriate play space for children in the interwar years and beyond.

Charles Wicksteed (1847–1931) was not a landscape designer, pedagogue or public health campaigner. He spent much of his life running his own businesses: initially steam ploughing in Suffolk and then a manufacturing company in Kettering. He married in 1877 and appears to have been a devoted parent to his three children. He was active in the Kettering and Northamptonshire Liberal Party, but even his daughter Hilda, who penned an otherwise ardent and diplomatic biography, felt that 'his service on local bodies was not outstandingly successful'.[54] This was perhaps epitomised by his endorsement of unsuccessful attempts to create a Royal Jubilee People's Park in Kettering in the 1880s.[55]

The success of Wicksteed's manufacturing business fluctuated in line with wider economic circumstances, as well as the success or otherwise of his products and inventions. His Stamford Road Works in Kettering was established in 1876 and manufactured a variety of goods at different times, from machine-tools and bicycles to motorcar gearboxes.[56] A small shed at the Works that had been making 'strong and endurable' wooden children's toys was converted to munitions production during the First World War.[57] On the back of a period of business success in the early 1900s, Wicksteed sought to purchase some land on the edge of Kettering. In January 1914, he completed the purchase of the country estate that had previously been associated with Barton Seagrave Hall. As the public park and playground that he subsequently created have been the most obvious and accessible features since then, it is perhaps unsurprising that both contemporary and historical accounts have tended to focus on examining his motivations for creating a public open space.[58] But it seems likely that the creation of a park was, at least initially, only a small part of a wider scheme.

There was much local speculation about Wicksteed's motives for purchasing the Barton Seagrave land, with opinions split over whether he was being foolish, eccentric or calculating.[59] Charitable donations had long provided a form of tax relief and putting the Barton Seagrave land into the charitable Wicksteed Village Trust would have avoided payment of income tax on the money involved, particularly at a time of significant wartime tax increases.[60] However, his religious and political values, as well as familial experiences seem to have been important motivating factors too. He had a powerful sense of moral responsibility, inspired in part by his Unitarian religious beliefs. In his book *Bygone Days and Now: A Plea for Co-Operation between Labour, Brains and Capital*, Wicksteed expressed the view that 'the whole edifice of modern civilization would fall to the ground without a foundation of sound moral principle ... all scientific inventions may come to nought, or even bring about evil, without moral guidance and inspiration'.[61] As a long-time radical Liberal, he firmly believed in capitalism, but also felt that the freedom of a laissez-faire economy and the technology it generated needed to be underpinned by rigorous moral standards. He had secured reasonable financial resources through his business and had a keen sense of obligation to those less fortunate, something common to many philanthropists who had created rather than inherited their wealth. In addition, seeing his own children benefit from access to more open space may partly have motivated him too. He felt that his second son in particular benefitted significantly when they moved to a house with a garden for the first time and as a result had much more space to run about.[62]

Not long after the land purchase, Wicksteed commissioned John Gotch, prominent architect and fellow Kettering Liberal Club member, to prepare a plan for the site.[63] The design was completed by June 1914 and showed a number of new roads, paths and, at the centre of the scheme, The Park. It included playing grounds for cricket, football and hockey, tennis courts, a large lake, tea pavilion and sunken garden. The plan set aside space for nurturing plants in hothouses, but at this stage there were no dedicated places to specifically cultivate children's health and well-being. Work soon started on the creation of the park and the clearance of existing landscape features. A copse of trees was felled and 'over 3,000 roots and stumps of one sort and another were uprooted by the aid of steam engines and dynamite'.[64] As was typical of the time, the playing grounds were created by levelling the landscape.

The plan that Gotch prepared for Wicksteed also included 150 building plots which bordered the park on three sides. At first glance, this plan could be seen as a direct descendant of nineteenth-century urban park projects, where property development and the establishment of green space went hand in hand. In the 1820s, John Nash had set out to create an appropriately salubrious and green environment for the wealthy residents of the large villas and terraces that were an integral part of the Regent's Park scheme in west London. Twenty years later, James Pennethorne had hoped to replicate this approach further east at Victoria Park, using income from the sale of large houses on the park boundary to offset the cost of creating a public green space. The title of Gotch's drawing provides a more salient clue as to the underlying assumptions and values that were to shape the estate for the next twenty years; Wicksteed was setting out to create the Barton Seagrave Garden Suburb Estate.[65]

In the 1880s, Wicksteed had been inspired by the influential American economist Henry George and his book *Progress and Poverty* (1879). As a result, he became an active campaigner on the issue of land nationalisation and explored ways to make the economic benefits of land ownership more socially equitable. Wicksteed was a prominent member of the Land Nationalisation Society (LNS, 1882) and in 1885 had written *The Land for the People*, a detailed assessment and promotion of the economic measures necessary to make land nationalisation financially, and therefore politically, viable.[66] In 1892 he followed this with *Our Mother Earth*, a more mainstream appeal for land nationalisation, which apparently achieved a circulation of 100,000.[67] By the late 1890s, there were close links between the LNS and the fledgling Garden Cities Association (1899) which had been formed around a core of LNS members and made use of its office space and staff. Ebenezer Howard had established the Association as a way to bring about radical social reform, but the involvement of influential philanthropists such as George Cadbury and William Lever meant that it soon focused more narrowly on ameliorating the living conditions of the working poor by improving their housing. Despite Howard's view that garden suburbs were antithetical to garden city ideals, the Garden City Association increasingly embraced green suburbs and town planning more generally.[68] Similarly, Wicksteed engaged with emerging ideas about town planning and was one of the opening speakers, along with noted planner Patrick Abercrombie, at the 1918 Leeds Civic Society House and Town Planning Exhibition, although the archives do not reveal the content of his address.[69]

There were a number of similarities between Howard and Wicksteed. Both were from nonconformist backgrounds, opposed contemporary military conflicts and were radical Liberals for much of their lives. Howard had been able to put his garden city ideals into practice in 1903 at Letchworth and purchasing the Barton Seagrave estate gave Wicksteed an opportunity to do something similar. It is possible that Wicksteed visited Letchworth as he travelled extensively to visit business customers and other open spaces around the country. More compellingly, Joseph Hartley Wicksteed, Charles's nephew and co-trustee of the Wicksteed Village Trust, lived in Letchworth and was an active member of the local community there on the eve of the First World War.[70]

Robert Fishman's description of Ebenezer Howard and the garden city could apply equally to Charles Wicksteed and the playground: 'with the ingenuity and patience of an inventor putting together a useful new machine out of parts forged for other purposes, [he] created a coherent design for a new environment'.[71] Wicksteed used staff, technology and skills ostensibly associated with his manufacturing business to shape the park environment and in time the playground too. In the postwar economic slump, Wicksteed put his underemployed staff to work excavating the park's lake. Tube-bending machines were put to use manufacturing an increasingly wide range of playground equipment. He used his own inventiveness to design a bread-and-butter machine and a jet-injected hot water supply system that could deliver four thousand cups of tea a day, so that the increasing number of park visitors could be served refreshments in a timely manner. Furthermore, Wicksteed adopted aspects of the garden city ideal in a number of ways, attempting to combine the benefits of town and country that Howard had illustrated in *Garden Cities of To-morrow*.[72] In practice this meant combining the beauty of nature, in an appropriately curated form, with the social opportunities and technologies of modern life. At Barton Seagrave there would be modern housing with private gardens and space for motorcars. The park would combine a picturesque landscape, large lake and mature trees, with modern recreational facilities and state-of-the-art canteen technology.

In terms of governance and ownership, the Barton Seagrave estate land was entrusted to the Wicksteed Village Trust in 1916, just as Howard had advocated and others had implemented, including the Cadbury family and the Bournville Village Trust (1900). Wicksteed also set out to pay higher wages to his employees and charged lower rents for the innovative prefabricated 'concrete cottages' he designed for local workers.[73] He disposed of building plots on 999-year leases as Howard had suggested, even though pressure from commercial investors meant that this had not been possible at Letchworth. Adherence to the garden city ideals was not simply a short-term impulse and did not end when the park became the increasing focus of Wicksteed's attention. The minutes of Wicksteed Village Trust meetings suggest that the sale of land to facilitate the creation of the garden suburb continued well into the 1930s; between 1920 and 1935, almost every trustee meeting involved the approval of land sales to individuals and local builders.[74]

In other ways, the Wicksteed Village Trust took a different path to Howard's Garden City model. Unlike at Letchworth, there was no overarching architectural vision for the suburb. In addition, it would be hard to see the Wicksteed Village Trust as a model of the cooperative values that were a key element of Howard's early thinking. At one of the first trustee meetings, a resolution was passed which stated that future meetings were only necessary once per year as Wicksteed had full control of the Trust. Furthermore, the trustees comprised family members and company employees, while meeting minutes show that money and land moved back and forth between the Trust, the company and individual trustees. The Objects of the Trust also hinted at Wicksteed's broader interests. While partly established to ameliorate the living conditions of the working-classes, the Trust was also tasked with preventing cruelty to animals and opposing vaccination.

In *Garden Cities of To-morrow*, Howard had considered how schools and wider recreational facilities would be created and managed, but children's recreation in particular was not explicitly mentioned. The term playground is used a number of times, but generally refers to a space for recreation – as in 'cricket fields, lawn-tennis courts, and other playgrounds' – rather than somewhere specifically for children.[75] Similarly, the Barton Seagrave Garden Suburb plan clearly showed large areas of parkland at the centre of the scheme and included 'playing grounds' for cricket, football, hockey and lawn tennis, but no dedicated space for children.

Excitement and freedom in Wicksteed Park

While the creation of the garden suburb continued well into the 1930s, the public profile of the park grew in prominence once the lake was completed in 1920. Local community organisations came together the following year to offer a tribute to Wicksteed, as a sign of public appreciation for the time and money he had invested in creating the park and lake. He explained in his acceptance speech that the initial impulse for the creation of play opportunities for children was accidental, rather than deliberate. 'Primitive' swings had been put up to coincide with a Sunday-school outing to the park, made of larch poles and chains. They proved so popular that he felt compelled to make them permanent and to provide more.[76] By 1923, a whole hockey pitch had been repurposed as a space to accommodate a remarkable number of 'play things', including sixty-two swings, fourteen see-saws and eight slides (Figures 2.3 and 2.4).

Figure 2.3: Wooden slides, c.1920, Wicksteed Park Archive, PHO-1614-4.

Figure 2.4: Large swings, c.1920, Wicksteed Park Archive, PHO-1614-5.

This interest in children's leisure and wellbeing was not entirely new. In addition to his earlier production of children's toys, Wicksteed's wider family were also active in campaigns to improve the lives of poor urban children and to provide more progressive educational opportunities. Wicksteed's older brother, Philip Henry Wicksteed, was a Unitarian minister, leading member of the Labour Church movement and a noted economist who produced one of the first critiques of Marx's theories in English.[77] However, it was Philip's role in the University Hall settlement where he encountered and sought to improve urban childhood.[78] The wider settlement movement had started in the 1880s and brought university graduates to poor urban areas to take part in voluntary social work, often with an emphasis on observing and organising children's leisure activities.[79] Furthermore, by the 1920s, Joseph Hartley Wicksteed had moved from Letchworth to London and was headteacher at the progressive King Alfred's School in Hampstead Garden Suburb, where considerable emphasis was placed on outdoor learning and individual freedom for pupils.[80] The wider impact of such progressive approaches to education on the children's playground will be explored in more detail in the next chapter, but it seems likely that Charles Wicksteed would have been exposed to some of these ideas through his family connections. In practice, he certainly embodied some of the values associated with child study that had emerged from the settlement movement, even if there is no surviving evidence of direct links with the British Child Study Association or its key international proponents, such as G. Stanley Hall or Maria Montessori. Wicksteed seems to have been one of the enthusiastic amateurs who rallied to the Association's cause to better understand the nature of childhood through observation.[81] He watched children playing in Wicksteed Park and noted how they liked to play. He visited other parks to see how children used them, but invariably found little of interest, except for old-fashioned swings, dangerous giant strides and clumsy see-saws.[82] No records remain that show where he went on his travels, but it seems likely that Wicksteed was visiting levelled and gravelled playgrounds and noting the type of gymnastic apparatus that had earlier been promoted by the MPGA and others.

Wicksteed's disappointment at the spaces that he visited led him to propose an alternative vision for the playground. Writing in a number of pamphlets and catalogues during the late 1920s, Wicksteed set out his own version of the playground ideal. His firm belief in personal freedom seems to have strongly influenced his attitudes toward children and the play space and equipment he created. Perhaps the reason he found little of interest when visiting other spaces was because the prescriptive nature of gymnastic equipment would have been at odds with his emerging notion of a play space as somewhere that should promote individual autonomy and enjoyment. He reflected that: ‘the poor little gutter-children with all their hardships, playing with mud in freedom, are far happier than many well-to-do children under the perfect control and sad dullness and weariness of a too-much-ordered life’.[83] Freedom for children, rather than regulation, would be a consistent feature of his playground rhetoric and action.

At the same time, Wicksteed embraced established ideas about the creation and management of public parks and playgrounds. Just like other campaigners, he felt strongly that the street was an inappropriate place for play, that green space could have a refining influence and that investment in the next generation would reap future benefits for society. He also emphasised the threat from hooligans and the importance of safety. Wicksteed & Co. were able to state that all of their products had been tested and refined in the Wicksteed Park playground before being put on the market, although this was not promoted to the unsuspecting visitors to Wicksteed Park.[84] The latest technology would avoid entanglement, deter over-swinging and prevent bumps and collisions, providing a safer playground experience. Wicksteed concluded that ‘it has been my policy if anything is not safe and unbreakable to make it so, or cease to use it’, encapsulating in one sentence the possibility that children could break play equipment and that play equipment could break children.[85]

It seems that contemporary play theories also informed Wicksteed’s thinking, to an extent at least. In particular, his writing suggests that he understood play as a way for children to expend surplus energy and direct their physical development. A playground would fulfil children’s natural urges to run, jump and play, as well as helping them to develop healthy bodies. In addition, open-air play would develop healthy tastes and a good temper, contributing to the appropriate development of their minds too. In some ways he would have agreed with Walter Wood’s claim in *Children’s Play* (1913) that municipal playgrounds could provide a healthy antidote to the unnatural urban environment.[86] At the same time, he would have disagreed strongly with Wood’s assertions that play spaces needed expert supervision and that girls and boys needed segregated play space due to inherent biological differences.

Instead, Wicksteed created a playground that was not physically segregated by gender and all children were instead encouraged to play together. While this was undoubtedly a progressive approach to play provision, Wicksteed was unlikely to be the first to have promoted or created shared play spaces. As early as 1915, the LCC’s park regulations listed thirty-one open spaces with facilities for children, but unlike earlier editions of the rules this version did not specify that facilities were segregated by gender.[87] While the revised regulations do not necessarily reflect changes to play spaces on the ground in London, they do suggest that attitudes towards prescribing specific areas for girls and boys had started to change, something that Wicksteed put into practice in Kettering. In addition, Wicksteed's view that supervision was unnecessary was based on the idea that children needed more activities, fewer regulations and a more prominent location for the playground in the park. He felt that people in general, and children in particular, 'want something doing' and not just spaces for genteel strolls or bucolic vistas. He argued that if play spaces were 'sufficient', in other words they provided enough things to do, then children would invariably get on better without an official attendant. Supervision was also unnecessary if play spaces were located in prominent locations. He argued that 'the Play Ground should not be put in a corner behind railings, but in a conspicuous and beautiful part of a Park, free to all, where people can enjoy the play and charming scenery at the same time; where mothers can sit, while they are looking on and caring for their children'.[88]

The idea of a separate domestic sphere for women, which included responsibility for raising children, would have seemed entirely natural for many Victorians, most likely including Wicksteed too. While he soon dispensed with the idea that girls and boys needed separate spaces to play, his attitude to women's place in the park and playground was more ambiguous.[89] In his earlier writing, Wicksteed focused on the benefits that a playground could offer mothers, presumably reflecting his personal experiences as well as contemporary ideas about the division of labour within families. Tea in the park canteen needed to be affordable so that housewives and their children could spend the day in the park for the least possible expense. Seating for mothers was 'very useful and a necessary adjunct to a play ground'.[90] But even here his views changed over time. By 1928 he felt it important that everyone should be admitted to his playground and he pondered 'why should you not let the father and mother come with the children of any age and enjoy the afternoon?'[91] There is evidence that Wicksteed created opportunities for women to participate in leisure activities, through a wartime Wicksteed & Co. women's football team and indirectly through the provision of a wide range of leisure facilities in Wicksteed Park.[92] In addition, the lack of regulations meant that in theory women could make use of all the facilities that had been provided. But his significance in this regard should not be over-emphasised and wider social norms continued to limit leisure opportunities for many women.

Historian Claire Langhamer has argued that age was a significant factor in women's access to leisure opportunities during the interwar period and evidence from Wicksteed Park and playground lends weight to this argument.[93] Generally, Wicksteed reinforced the notion that women who had children should primarily occupy the domestic sphere. He emphasised women's domestic responsibilities as mothers, promoted the need for seating in the playground so that they could supervise their children, and provided affordable refreshments to make catering for their family easier. In a way, the Wicksteed playground could be seen as an extension of domestic life, a place where women were expected to continue fulfilling their domestic duties, supervising children and providing sustenance. However, the playground also potentially disrupted patterns of domesticity. Where children had traditionally played in the street within calling distance of home, supervising children at play had invariably been an informal, sociable and collective endeavour for working-class women.[94] In contrast, if the children of Kettering were encouraged to play in the Wicksteed Park playground, then mothers were expected to come too, potentially interrupting established patterns of social support and community life. Wicksteed imagined that a trip to the park provided a holiday for mothers and their children, but at the same time it created an expectation that mothers were responsible for transporting their children to a place where they could play, as well as directly supervising them while there.

The provision of affordable refreshments may have made a family visit to the playground easier, but it was also part of a wider attempt to make the park financially sustainable. Unlike municipal open spaces, Wicksteed Park did not have access to state funding. Wicksteed had considered the longer-term financial viability of the Wicksteed Village Trust from early on, but the difficulties it faced are evident from its annual accounts. One thousand two hundred Wicksteed & Co. shares were given to the Trust in 1920 to provide an ongoing source of income. Dividends were small and from 1916 to the early 1930s rental income and farm sales (including potatoes, wheat, oats and turf) far exceeded income from investments and park-related activities, including boat hire and the sale of refreshments. Despite a diverse range of income sources, the Trust spent far more than it earned and by 1931 had total debts of £4.7m (£23,452).[95] The sale of souvenir booklets was one of the ways in which the Trust attempted to generate income to reinvest in the park, but they also provide an insight into the way that the park and its landscape were presented to visitors.

The geographer David Matless has argued that both tradition and modernity were key characteristics of interwar conceptions of the rural landscape and the moral geographies imprinted on it.[96] Matless finds that these values were significant in organisations such as the Ramblers, the

Youth Hostel Association and the Council for the Preservation of Rural England – but they also influenced the way that Wicksteed Park and children's place in it were presented to visitors. In a souvenir booklet from the 1920s, an image of the strikingly modern park pavilion is combined with classical statues and urns. The sandpit in particular and the playground more generally are presented as bustling places where girls and boys play together, while more bucolic images of the wider park landscape bear similarities with the picturesque grounds of the country estate.[97] In a 1936 souvenir, aerial photography showcased park features from a novel, modern perspective, while at the same time emphasising the park's rural surroundings.[98] This combination of modern urbanity and traditional rurality were even incorporated into the headed paper of the Trust: in the foreground are the sandpit, playground, pavilion and people, while in the background there are fields, hedgerows and trees all the way to the horizon.[99]

As such, the park was not presented primarily as a retreat from the modern world as it had been in the earlier rhetoric of park advocates. Instead, Wicksteed Park built on a long tradition of manufactured items and industrial materials in park landscapes. It was presented as a place to engage with the benefits of modernity, including the use of engineering technology that promoted exciting leisure activities for both children and adults, but at the same time was framed by a rural backdrop, with the rolling landscape and mature trees providing a health-inducing dose of nature. In particular, the technological modernity of the amusement park landscape was increasingly influential on the form of Wicksteed Park. Perhaps the most notable example was the water chute, designed and installed by Wicksteed in 1926 and now Grade-II listed, but the installation of a miniature railway in the 1930s also reflected the influence of commercial amusement rides on the park landscape.[100] In the playground specifically, Wicksteed's Joy Wheel seems to have been directly inspired by the similarly named mechanical roundabouts that were used at Great Yarmouth and Blackpool from around 1913.[101] But while the fairground version was mechanised and attempted to displace its riders using high speed and centrifugal force for the amusement of onlookers, Wicksteed's £7,000 (£40) version was self-propelled, smaller and included plenty of places to grip on.

The Ocean Wave, which cost £5,600 (£30), also appears to have been inspired by a circus ride. According to *The Times*, Hengler's Circus in London installed an Ocean Wave for the first time in Britain in 1890. Inspired by a similar ride seen in Paris, it could accommodate over one hundred passengers and mimicked the motion of a sailing boat.[102] But where the circus version had a circumference of 55m (180 ft) and included six small yachts, the playground version was less than half the size at 22m (75ft) and children stood or sat directly on the metal framework.[103] Although reduced in scale, such equipment and the souvenir booklets in which it was represented embodied notions of excitement and adrenaline, as well as the health benefits of a bucolic parkland setting. This combination of amusement-style rides and green landscape further highlights the complexities of early twentieth-century rural modernism that have been a feature of scholarship on interwar film, architecture and infrastructure.[104] But Wicksteed also challenged the traditional conceptions of the playground, as excitement replaced structured forms of exercise as the rationale for play space form.

Wicksteed also challenged established notions of appropriate park behaviour. In sharp contrast to the earlier attempts at regulation, the Wicksteed Park souvenirs emphasised an alternative attitude to park users and their conduct. One booklet enthused that

> one of the charms of the place is perhaps the freedom that is everywhere. There are no notices to keep off the grass, or not to do anything else. All go into the park to do what they like and to go where they like. It has had a distinctly civilizing influence and gives much health and happiness. The freedom granted is seldom abused.[105]

Facilities were provided in the park to support children's autonomy in exploring both the park environment and their individual abilities. Children and adults were not only welcome to paddle in the lake but also to fall in and get soaked. A nurse attendant would help anyone that fell in by providing a temporary change of clothes, while their wet garments were quickly dried in a specially designed hot air cabinet.[106] Children's playful activities were not frowned upon, nor constrained by regulations and railings. Instead Wicksteed set out to help mitigate the consequences of playfulness, rather than attempting to regulate and control it. In a similar way, Wicksteed set out to design and build play equipment that was strong enough to withstand the myriad ways that both children and adults would use it, rather than attempting to adjust users' behaviour to accommodate the technical constraints of the equipment. Wicksteed repeatedly emphasised his sentiment that it was 'easier for me, as an engineer, to make a swing strong enough to hold all who come than to keep park-keepers bawling at the youths all day long'.[107] In naming his new products, he also tended to focus on monikers that emphasised the playful nature of the playground – the Joy Wheel and Jazz Swing being early examples.

In practice, freedom was not absolute and Wicksteed was quick to express his displeasure at what he felt was inappropriate behaviour. Just as Matless has shown for the interwar countryside, objections to littering were a key component of the moral geography of Wicksteed Park. According to Wicksteed, littering disfigured the landscape and offended his idea of good citizenship. In a letter to the local paper, he emphasised the personal distress caused to him by both the littering and the potential need to increase the cost of a jug of tea to cover the wages of an additional attendant to pick up the litter.[108] Freedom came with individual responsibility, mirroring wider social processes that linked public parks and other green spaces with the construction of appropriate forms of citizenship.[109]

The form of citizenship promoted at Wicksteed Park had much less to do with creating colonial identities than was the case in other parks, where architecture and pageants sought to instil the values of empire.[110] Perhaps Wicksteed shared with his Liberal colleague Charles Masterman a sense that the empire represented a force for national self-indulgence rather than greatness, while Wicksteed certainly deplored the failures of statecraft that he felt resulted in the First World War.[111] Moreover, he imagined the playground as a space of enjoyment and freedom for individual children and their families, inspired by the healthiness of green space and the benefits of entertaining physical movement, rather than the prescriptions of the children's gymnasium and its geopolitical assumptions. But at the same time Wicksteed was not averse to taking advantage of the business opportunities that both the First World War and empire created. During the war, his Stanford Road factory was converted to munitions production and in the 1920s and 1930s Wicksteed & Co. were able to take advantage of the commercial opportunities provided by imperial networks. Soon after Wicksteed Park opened, politicians from the parishes around Kettering saw the playground and requested similar facilities for their local communities.[112] Over time Wicksteed & Co. went on to equip thousands of playgrounds across the UK and beyond. In a 1936 advert the company claimed to have supplied over 3,000 playgrounds with their equipment, a figure that had increased to 4,000 only a year later.[113] In the 1920s, Wicksteed & Co. exported playground equipment to South Africa. By the 1950s, a postwar export drive saw them provide equipment for play spaces in Canada, New Zealand, India, Hong Kong, Malta, the West Indies, North Borneo, Southern Rhodesia and St Helena, as well as the Belgian Congo, Venezuela and the USA.[114]

In establishing the children's playground as an export template, Wicksteed & Co. contributed towards the increasing standardisation of the playground form. For example, the equipment sold to city authorities in South Africa was identical to that sold in Britain. The Joy Wheel pictured in Joubert Park, Johannesburg, is the same model displayed in Wicksteed's brochures from the 1920s. While the playground form may have become increasingly standardised as a result, Wicksteed's vision of the playground as a space where all children had freedom to play did not become the guiding principle of playground management. Instead, local cultural values shaped the way that children experienced playground spaces. For example, the provision of playgrounds in Johannesburg was likely to have been part of the city's longstanding connection with the developments and cultural styles of other international cities, including London and New York, and Johannesburg Council's concerted attempts at modernisation in the 1920s.[115] Along with other imported trends, including modernist high-rise buildings, new retail stores and swimming pools, the playground was one expression of the enduring connection between the city's white, middle-class councillors and British ideas and values. The creation of playground spaces also reinforced racial, cultural and class segregation. Few facilities were built in black neighbourhoods and black children were allowed to use the playground in Joubert Park just once a year.[116] This tentative exposure of the connections between Kettering and Johannesburg and the racial politics of the playground undoubtedly demands further research and there is evidence that archive material exists elsewhere to inform a broader study, including in Cape Town.[117]

In the first two decades of the twentieth century, the social, political and environmental problems of the city remained a powerful justification in the minds of advocates for greater playground provision. The ideal physical form of such spaces was far from settled, with levelled playgrounds, children's gymnasiums and the US-inspired organised playground all in circulation. The First World War increased anxiety about children's place in society and there were renewed calls for the provision of amenities for children, partly to promote positive behaviour but also to enhance children's physical and mental wellbeing, especially close to home. An increasing number of commercial equipment suppliers offered products that reinforced normative assumptions about age, gender and exercise. But these prescriptive assumptions were challenged by Charles Wicksteed in the park he created and through the products that his company manufactured and sold. Rather than a bolster for wider imperial ambitions, Wicksteed imagined the playground as a space of excitement and freedom, with the equipment he created inspired by the fairground and amusement park, rather than solely the gymnasium. He also reacted against the segregation of play spaces by age and gender and instead created a playground in Wicksteed Park that children and adults were welcome to use together. The quantity of land available to Wicksteed at Barton Seagrave meant that playground technology and industrial materials could be situated within extensive green landscapes, combining aspects of traditional park rhetoric with the modernity of the amusement park, providing the benefits of both town and country. With Wicksteed Park as a proven testing ground, Wicksteed & Co. were able to sell their products in increasing numbers, along with a persuasive vision for the children's playground, but only some of the values that shaped the management of Wicksteed Park travelled with these products. In Britain, the segregation of spaces for children by gender became increasingly uncommon, but elsewhere local social and cultural values shaped access to the playground. Just as others were adopting the standard Wicksteed version of the playground, with its swings, slides and roundabouts, the trustees of Wicksteed Park had to identify new sources of income, initiating a shift in emphasis away from the free-to-use playground and towards income-generating rides that characterise the park today.

Notes

1. Henry S. Curtis, *Education Through Play* (New York: Macmillan, 1915), p. 138.

2. Alicia Amherst, *London Parks and Gardens* (London: Archibald Constable & Co., 1907); Jason Tomes, 'Amherst, Alicia Margaret, Lady Rockley (1865–1941), Garden Historian', *Oxford Dictionary of National Biography* (Oxford: Oxford University Press, 2004).

3. Annesley Kenealy, 'Playgrounds in the Parks: A Plea for the Children', *The Daily Mail*, 14 March 1907, p. 6.

4. 'Children in Kensington Gardens', *The Times*, 7 August 1909, p. 6.

5. A.J. Greene, 'The Open Air School Movement, 1904–1912', *The Public Health Journal*, 3.10 (1912), 547–52; David Pomfret, 'The City of Evil and the Great Outdoors: The Modern Health Movement and the Urban Young, 1918–40', *Urban History*, 28 (2001), 405–27.

6. Clare Hickman, 'Care in the Countryside: The Theory and Practice of Therapeutic Landscapes in the Early Twentieth Century', in *Gardens and Green Spaces in the West Midlands since 1700*, ed. Malcolm Dick and Elaine Mitchell (Hatfield: University of Hertfordshire Press, 2018), pp. 160–85.

7. Helen Pussard, 'Historicising the Spaces of Leisure: Open-Air Swimming and the Lido Movement in England', *World Leisure Journal*, 49.4 (2007), 178–88.

8. 'Physical Recreation, Manchester and Playing Fields', *The Manchester Guardian*, 7 November 1913, p. 16.

9. Charles Masterman, 'Realities at Home', in *The Heart of Empire: Discussions of Problems of Modern City Life in England, with an Essay on Imperialism*, ed. Charles Masterman (London: Fisher Unwin, 1902), pp. 1–52 (p. 3); H.C.G. Matthew, 'Masterman, Charles Frederick Gurney (1873–1927), Politician and Author', *Oxford Dictionary of National Biography* (Oxford: Oxford University Press, 2015).

10. Royal Parks, 'Kensington Gardens Children's Playground', 1909, National Archives, WORKS/16/391.

11. Royal Parks, 'Primrose Hill Children's Playground, Gymnasium and Lavatories', 1938, National Archives, WORKS/16/1670.

12. Ian Beckett, 'The Nation in Arms, 1914–1918', in *A Nation in Arms: A Social Study of the British Army in the First World War*, ed. Ian Beckett and Keith Simpson (Barnsley: Pen and Sword, 2004), pp. 1–36.

13. 'Juvenile Crime', *The Times*, 13 May 1916, p. 3.

14. Robert Snape, 'Juvenile Organizations Committees and the State Regulation of Youth Leisure in Britain, 1916–1939', *Journal of the History of Childhood and Youth*, 13.2 (2020), 247–67.

15. 'Young Offenders in War Time', *The Times*, 13 March 1916, p. 5.

16. Margaret Hogarth, 'Campbell, Dame Janet Mary (1877–1954), Medical Officer', *Oxford Dictionary of National Biography* (Oxford: Oxford University Press, 2006).

17. Janet M. Campbell, *The Physical Welfare of Mothers and Children* (Dunfermline: Carnegie United Kingdom Trust, 1917).

18. W. Leslie Mackenzie, *Scottish Mothers and Children* (Dunfermline: Carnegie United Kingdom Trust, 1917), pp. 335–7.

19. *The Leisure of the People: A Handbook, Being the Report of the National Conference Held at Manchester, November 17th–20th, 1919*. (Manchester: Conference Committee, 1919), p. 45.

20. Robert Snape and Helen Pussard, 'Theorisations of Leisure in Inter-War Britain', *Leisure Studies*, 32 (2013), 1–18.

21. Peter Borsay, *A History of Leisure: The British Experience since 1500* (Basingstoke: Palgrave Macmillan, 2006).

22. Stephen Jones, *Workers at Play: A Social and Economic History of Leisure 1918–39* (London: Routledge, 1986), p. 93 The proportional difference between historical and present-day amounts here is due to the impact of economic deflation in the intervening years.

23. Thomas Richards, *The Commodity Culture of Victorian England: Advertising and Spectacle, 1851–1914* (London: Verso, 1991).

24. Claire L. Jones, *The Medical Trade Catalogue in Britain, 1870–1914* (London: University of Pittsburgh Press, 2013).

25. Royal Parks, 'Children's Playgrounds: Gifts of Equipment Offers and Acceptances', 1923, National Archives, WORKS/16/1705.

26. F.H.W. Sheppard, 'The Crown Estate in Kensington Palace Gardens: Individual Buildings', in *Survey of London*, Northern Kensington (London: London County Council, 1973), XXXVII, 162–93 http://www.british-history.ac.uk/survey-london/vol37/pp162-193 [accessed 16 November 2023].

27. Spencer, Heath and George Ltd, 'Playground Catalogue', 1927, National Archives, WORKS/16/1705; Bayliss, Jones and Bayliss Ltd, 'Gymnasia for Parks and Recreation Grounds, School Playgrounds, Etc.', 1912, National Archives, WORKS/16/1705.

28. Spencer, Heath and George Ltd to D. Campbell, 'Sketch for Regent's Park Superintendent', 1 May 1924, National Archives, WORKS/16/1705.

29. Spencer, Heath and George Ltd, 'Playground Catalogue'.

30. Bayliss, Jones and Bayliss Ltd, 'Gymnasia for Parks'.

31. Geoffrey Pearson, '"A Jekyll in the Classroom, a Hyde in the Street": Queen Victoria's Hooligans', in *Crime and the City*, ed. David Downes (London: Macmillan, 1989), pp. 10–35.

32. Bayliss, Jones and Bayliss Ltd, 'Gymnasia for Parks', p. 16.

33. 'See-Saw, n. and Adj.', *Oxford English Dictionary* (Oxford: Oxford University Press, 2006).

34. Curtis, *Education Through Play*, p. 143; Charles Wicksteed & Co., 'Playground Equipment, Tennis Posts, Fencing and Park Seats', 1926, Wicksteed Park Archive, uncatalogued.

35. Ralph Harrington, 'On the Tracks of Trauma: Railway Spine Reconsidered', *Social History of Medicine*, 16.2 (2003), 209–23.

36. Spencer, Heath and George Ltd, 'Sketches of Regulation Swing Frame, Patent Safety Giant Stride & See-Saw', n.d., National Archives, WORKS/16/1705.

37. E. Garrity, 'Letter to Chief Commissioner of Works Regarding Indecent Offences towards Children Metropolitan Radical Federation', 22 July 1913, National Archives, WORKS/16/532.

38. Deputy Education Officer to Board of Works, 'Inadequate Supervision of Open Spaces', 17 June 1914, National Archives, WORKS/16/532.

39. Royal Parks, 'Prevention of Offences against Children (Various Minutes and Notes)', 1913, National Archives, WORKS/16/532.

40. Elizabeth Gagen, 'Playing the Part: Performing Gender in America's Playgrounds', in *Children's Geographies: Playing, Living, Learning*, ed. Sarah Holloway and Gill Valentine (London: Routledge, 2000), pp. 213–29; Elizabeth Gagen,

'An Example to Us All: Child Development and Identity Construction in Early 20th-Century Playgrounds', *Environment and Planning A: Economy and Space*, 32.4 (2000), 599–616.

41. London County Council, 'Regulations Relating to the Playing of Games at Parks and Open Spaces under the Control of the Council', 1904, p. 12, LSE Library, 421 (129A).

42. Bayliss, Jones and Bayliss Ltd, 'Gymnasia for Parks', p. 16.

43. J.C. Loudon, *An Encyclopædia of Gardening* (London: Longman, Rees, Orme, Brown, Green and Longman, 1835), pp. 40, 109, 158, 229.

44. Jeffrey A. Auerbach, *The Great Exhibition of 1851: A Nation on Display* (New Haven: Yale University Press, 1999), p. 105.

45. Jonathan Conlin, 'Vauxhall Revisited: The Afterlife of a London Pleasure Garden, 1770–1859', *Journal of British Studies*, 45 (2006), 718–43; Jonathan Conlin, 'Vauxhall on the Boulevard: Pleasure Gardens in London and Paris, 1764–1784', *Urban History*, 35 (2008), 24–47.

46. Peter Borsay, 'Pleasure Gardens and Urban Culture in the Long Eighteenth Century', in *The Pleasure Garden, from Vauxhall To Coney Island*, ed. Jonathan Conlin (Philadelphia: University of Pennsylvania Press, 2013), pp. 49–77.

47. 'The Showman World', *The Era*, 6 May 1899, p. 18.

48. Thomas Murphy, 'The Evolution of Amusement Machines', *Journal of the Royal Society of Arts*, 99.4855 (1951), 791–806.

49. Gary Cross and John K. Walton, *The Playful Crowd: Pleasure Places in the Twentieth Century* (New York: Columbia University Press, 2005).

50. P.G. Wodehouse, *Love among the Chickens* (New York: Circle, 1909), p. 238.

51. Josephine Kane, *The Architecture of Pleasure: British Amusement Parks 1900–1939* (Farnham: Ashgate, 2013).

52. Robert Snape, 'The New Leisure, Voluntarism and Social Reconstruction in Inter-War Britain', *Contemporary British History*, 29 (2015), 51–83.

53. Rhodri Davies, *Public Good by Private Means* (London: Alliance Publishing Trust, 2015), p. 134; National Playing Fields Association, *Second Annual Report* (National Playing Fields Association, 1928), p. 13, National Archives, CB 4/1.

54. Hilda M. Wicksteed, *Charles Wicksteed* (London: J.M. Dent & Sons, 1933), pp. 70, 85–7.

55. Ian Addis, *Out to Play in Kettering* (Kettering: Bowden Publications, 2013), p. 5.

56. The Wicksteed Park Archive holds a number of uncatalogued papers including patent drawings, advertisements, instruction booklets.

57. Hilda M. Wicksteed, *Charles Wicksteed*, p. 96.

58. Hilda M. Wicksteed, *Charles Wicksteed*, p. 101.

59. 'Wicksteed Park: Kettering Clubmen's Appreciation of the Founder', *The Kettering Leader*, 15 July 1921, p. 7.

60. Davies, *Public Good by Private Means*, p. 109; M.J. Daunton, 'How to Pay for the War: State, Society and Taxation in Britain, 1917–24', *The English Historical Review*, 111 (1996), 882–919.

61. Charles Wicksteed, *Bygone Days and Now: A Plea for Co-Operation between Labour, Brains and Capital* (London: Williams & Northgate, 1929), p. 150.

62. Charles Wicksteed, *A Plea for Children's Recreation after School Hours and after School Age* (Kettering: Wicksteed Charitable Trust, 1928), p. 10.

63. Gotch also designed Charles Wicksteed's home, Bryn Hafod, in 1898 and would later become President of the Royal Institute of British Architects from 1923 to 1925. Ian MacAlister and John Elliott, 'Gotch, John Alfred (1852–1942)', *Oxford Dictionary of National Biography* (Oxford: Oxford University Press, 2004).

64. Wicksteed Village Trust, 'An Account of the Wicksteed Park and Trust', 1936, p. 7, Wicksteed Park Archive, BRC-1906.

65. Gotch & Saunders, 'Barton Seagrave Garden Suburb Estate', 1914, Wicksteed Park Archive, uncatalogued.

66. Charles Wicksteed, *The Land for the People: How to Obtain It and How to Manage It* (London: William Reeves, 1885); Land Nationalisation Society, *Report 1885–6* (London: Land Nationalisation Society, 1886).

67. Charles Wicksteed, *Our Mother Earth: A Short Statement of the Case for Land Nationalisation* (London: Swan Sonnenschein & Co, 1892); Hilda M. Wicksteed, *Charles Wicksteed*, p. 68; Charles later withdrew from the movement as it increasingly advocated industrial nationalisation too, something he opposed – see Charles Wicksteed, 'National Coal: The Farce of Nationalisation Exposed', n.d., Wicksteed Park Archive, MGA-3006.

68. Robert Beevers, *The Garden City Utopia: A Critical Biography of Ebenezer Howard* (Basingstoke: Macmillan, 1988), pp. 37, 71, 133.

69. Leeds Civic Society, 'House and Town Planning Exhibition Programme', 1918, Wicksteed Park Archive, PRG-3004.

70. National Union of Women's Suffrage Societies, 'Letchworth and District Society', 1912, Garden City Collection, LBM2988; Letchworth Dramatic Society, 'A Variety Entertainment', 1914, Garden City Collection, LBM4007.18; W.H.G. Armytage, *Heavens below: Utopian Experiments in England 1560–1960* (London: Routledge and Kegan Paul, 1961), p. 398.

71. Robert Fishman, *Urban Utopias in the Twentieth Century: Ebenezer Howard, Frank Lloyd Wright, Le Corbusier* (Cambridge, MA: MIT Press, 1982), p. 28.

72. Ebenezer Howard, *Garden Cities of To-Morrow* (London: Swan Sonnenschein & Co, 1902).

73. Charles Wicksteed, 'Concrete Cottages', *The Machine Tool Review*, 1920, Wicksteed Park Archive, uncatalogued.

74. Wicksteed Village Trust, 'Minute Book, 1920–1935', Wicksteed Park Archive, uncatalogued.

75. Howard, *Garden Cities of To-Morrow*, p. 63.

76. 'Mr. Chas. Wicksteed's Generosity: Kettering Club's Appreciation, Mr Wicksteed Silences His Critics', *The Kettering Guardian*, 15 July 1921, p. 6.

77. Ian Steedman, 'Wicksteed, Philip Henry (1844–1927), Unitarian Minister and Economist', *Oxford Dictionary of National Biography* (Oxford: Oxford University Press, 2004); Krista Cowman, '"A Peculiarly English Institution": Work, Rest, and Play in the Labour Church', *Studies in Church History*, 37 (2002), 357–67.

78. University Hall Settlement, 'Memorandum and Articles of Association' (London, 1895), p. 11, LSE Library, FOLIO FHV/G60.

79. Kate Bradley, 'Creating Local Elites: The University Settlement Movement, National Elites and Citizenship in London, 1884–1940', in *In Control of the City: Local Elites and the Dynamics of Urban Politics, 1800–1960*, ed. Stefan Couperus, Christianne Smit, and Dirk Jan Wolffram, Groningen Studies in Cultural Change, 28 (Leuven: Peeters, 2007), pp. 81–92.

80. W.A.C. Stewart, *Progressives and Radicals in English Education 1750–1970* (London: Macmillan, 1972).

81. Kevin Brehony, 'Transforming Theories of Childhood and Early Childhood Education: Child Study and the Empirical Assault on Froebelian Rationalism', *Paedagogica Historica*, 45 (2009), 585–604 (p. 595).

82. Charles Wicksteed, *A Plea for Children's Recreation*, pp. 6, 8.

83. Charles Wicksteed, *Bygone Days and Now*, p. 55.

84. Charles Wicksteed & Co. to City of Lincoln Surveyor, 'Children's Playground Equipment', 21 November 1933, Wicksteed Park Archive, uncatalogued; 'Swedish-Inspired Play Equipment Experiment', *Evening Telegraph*, 29 July 1964, p. 6, Wicksteed Park Archive, uncatalogued.

85. Charles Wicksteed, *A Plea for Children's Recreation*, p. 7.

86. Walter Wood, *Children's Play and Its Place in Education* (London: Kegan Paul, Trench, Trübner & Co., 1913), p. 179.

87. London County Council, 'Parks and Open Spaces: Regulations Relating to Games, Together with Particulars of the Facilities Afforded for General Recreation', 1915, LSE Library, 421 (129D).

88. Charles Wicksteed & Co., 'Playground Equipment', 1926, p. 6.

89. Charles Wicksteed, *A Plea for Children's Recreation*, p. 6.

90. Charles Wicksteed & Co., 'Playground Equipment', 1926, p. 7.

91. Charles Wicksteed, *A Plea for Children's Recreation*, p. 24.

92. 'Photo of Wicksteed's Munition Girls F.C. Team', n.d., Wicksteed Park Archive, uncatalogued.

93. Claire Langhamer, *Women's Leisure in England 1920–60* (Manchester: Manchester University Press, 2000).

94. Krista Cowman, 'Play Streets: Women, Children and the Problem of Urban Traffic, 1930–1970', *Social History*, 42 (2017), 233–56 (p. 237).

95. Charles Wicksteed & Co., 'Director Minute Book, 1920–1956', Wicksteed Park Archive, uncatalogued; Wicksteed Village Trust, 'Minute Book, 1920–1935'; Wicksteed Village Trust, 'Annual Accounts 1916–48', Wicksteed Park Archive, uncatalogued.

96. David Matless, *Landscape and Englishness* (London: Reaktion Books, 1998).

97. Wicksteed Village Trust, 'The Wicksteed Park Souvenir', n.d., London Metropolitan Archive, CLC/011/MS22290.

98. Wicksteed Village Trust, 'An Account of the Wicksteed Park and Trust'.

99. Wicksteed Village Trust to J. Brandon-Jones, Letter, 4 October 1946, Wicksteed Park Archive, LET-1044.

100. Historic England, *Water Chute at Wicksteed Park, National Heritage List for England, 1437706*, 2016.

101. Kane, *The Architecture of Pleasure*, p. 55.

102. 'The Ocean Wave', *The Times*, 3 January 1890, p. 4.

103. Charles Wicksteed & Co., 'Playground Equipment', 1926.

104. Kristin Bluemel and Michael McCluskey, eds., *Rural Modernity in Britain: A Critical Intervention* (Edinburgh: Edinburgh University Press, 2018).

105. Wicksteed Village Trust, 'The Wicksteed Park Souvenir', p. 1.

106. Wicksteed Village Trust, 'The Wicksteed Park, Kettering', n.d., Wicksteed Park Archive, BRC-1199.

107. Charles Wicksteed, *A Plea for Children's Recreation*, p. 15; Charles Wicksteed & Co., 'Playground Equipment', 1926, p. 7.

108. Charles Wicksteed, 'The Pity of It: Thoughtless Picnic Parties in the Wicksteed Park', *The Kettering Leader*, 29 July 1921, p. 5.

109. Hazel Conway, 'Everyday Landscapes: Public Parks from 1930 to 2000', *Garden History*, 28 (2000), 117–34.

110. Ruth Colton, 'Savage Instincts, Civilising Spaces: The Child, the Empire and the Public Park, c.1880–1914', in *Children, Childhood and Youth in the British World*, ed. Shirleene Robinson and Simon Sleight (Basingstoke: Palgrave, 2016), pp. 255–70; Joanna Brück, 'Landscapes of Desire: Parks, Colonialism, and Identity in Victorian and Edwardian Ireland', *International Journal of Historical Archaeology*, 17 (2013), 196–223.

111. Matthew, 'Masterman, Charles Frederick Gurney (1873–1927)'; Hilda M. Wicksteed, *Charles Wicksteed*, p. 96.

112. Charles Wicksteed & Co., 'Play Things as Used in the Wicksteed Park', 1923, p. 3, National Archives, WORKS/16/1705.

113. 'Advert for Wicksteed & Co. Playground Equipment', *Journal of Park Administration, Horticulture and Recreation*, 1.1 (1936), 1 (p. 1); W.W. Pettigrew, *Municipal Parks: Layout, Management and Administration* (London: Journal of Park Administration, 1937), p. xiv, RHS Lindley, 969.2 PET.

114. 'The Children of St Helena', *Machinery Lloyd*, 29 (1957), 1–2.

115. Louis Grundlingh, 'Municipal Modernity: The Politics of Leisure and Johannesburg's Swimming Baths, 1920s to 1930s', *Urban History*, 49.4 (2022), 771–90.

116. Charles Wicksteed, *A Plea for Children's Recreation*, p. 23.

117. 'Playground Equipment (Charles Wicksteed)', 1927, Cape Town Archives Repository, 3/CT 4/1/4/71 B410/4.

Chapter 3

Playgrounds for the people: 'a magnetic force to draw children away from the dangers and excitements of the streets'

In the 1920s and 1930s, a combination of philanthropic funding, voluntary action and municipal enthusiasm saw children's playgrounds created in increasing numbers and greater consensus about the ideal playground form. Edward Prentice Mawson, the prominent landscape architect, suggested that prior to the First World War the intrusion of 'a children's playground into the parks was regarded as vandalism and was frequently the subject of bitter controversy'.[1] By the 1930s, children's playgrounds had become a relatively common feature of public parks and the design of these play spaces was dominated by manufactured equipment. Writing in 1937, the respected park superintendent and broadcaster W.W. Pettigrew summed up the state of play, so to speak, in pointing out that the broad principle of providing equipped playgrounds had been fully recognised.[2] In exploring this shift in attitudes, the National Playing Fields Association (NPFA) might seem like an unlikely advocate for children's playgrounds. While its name suggested a preoccupation with spaces for sports, it quickly became an important sponsor and source of expertise on playgrounds during the interwar period. It brought together existing campaigners and organisations to promote – and unintentionally standardise – the provision of playgrounds and playing fields in both urban and rural areas.

The entry of the NPFA into matters of children's recreation took place at a time of significant and complex social change after the First World War. In particular, the 'problems' of leisure, citizenship, gender and class concerned many contemporaries. The NPFA response was equally complex and initially combined nineteenth-century ideas about childhood, class and gender with twentieth-century attempts to provide suitable recreational spaces for the modern world. Its rhetoric often drew upon prewar notions of imperial masculinity and for a time continued to emphasise the physical degeneration of the urban working-classes and the moral dangers of the street. In doing so, it demonstrated an ongoing belief in the power of the built environment to shape individual and collective behaviour, even as medical thinking was increasingly sceptical of open-air treatments for illness. At the same time, the NPFA stressed the modernising potential of properly equipped playgrounds for existing and new communities in both rural and urban areas. Despite its 'national' moniker and standardising tendencies, the NPFA operated through local branches which funded and sought to influence the work of municipal authorities in their attempts to promote active, healthy outdoor recreation. As such, the NPFA serves as an example of the ongoing importance of municipal authorities and voluntary action in addressing social and environmental problems, and the interaction of national and local, urban and rural actors.[3]

This chapter plots the evolution of the NPFA and its endeavours in the field of children's play to explore the increasingly common provision of playgrounds and the development of an amenity standard in the first half of the twentieth century. It also uncovers the emerging tensions between advocates' emphasis on the playground as a site of safety, particularly as a response to the dangers associated with playing on the street among an increasing number of motor vehicles, and the real and imagined threats from apparatus, adults and animals.

Playing fields and playgrounds in interwar Britain

There had been local calls for the protection of existing playing fields and campaigns for the creation of new ones since the late nineteenth century. Organisations such as the London Playing Fields Committee (1890) and the Manchester and Salford Playing Fields Society (1907) had emphasised the ways in which playing fields could help to tackle physical degeneration and improve the character and morals of urban youths.[4] By 1924 there was a growing sense that these local efforts needed to be coordinated and expanded. Recalling nineteenth-century demands for parks for the people, a number of prominent politicians signed an open letter to the national press in April 1925 calling for 'playing fields for the people', spaces that were distinct from public parks, gardens and commons.[5] Signatories came from across the political spectrum and included government ministers, other high-profile politicians, as well as social reformers and campaigners. The letter argued that space for active recreation would contribute to both improved individual health and national efficiency and was therefore of significant domestic and imperial importance. Implicitly, their scheme suggested that participation in active recreation was a civic responsibility that needed to be performed across the nation by all sections of the community.[6]

There is little remaining evidence of the planning that went into the creation of the new organisation, but it would appear that playground advocates had a significant impact on its objectives, strategies and actions. An early draft of its constitution had a narrow typological focus on playing fields to facilitate participation in sport.[7] By the time the new organisation was formally launched as the National Playing Fields Association in July 1925, both children and playgrounds had become fundamental to its stated objectives. The first edition of its quarterly journal, *Playing Fields*, stated that the NPFA's two main objectives were to provide playgrounds for small children and playing fields for the masses.[8] At its formal inauguration at the Royal Albert Hall, senior politicians, royalty and celebrities were united in their support for the new organisation, while former prime minister David Lloyd George famously declared that 'the right to play is a child's first claim on the community'.[9]

Given the widespread political support, it is perhaps surprising that the state did not lead attempts to provide recreational facilities in the same way that it provided infrastructure for education and other social work. However, the provision of public parks and the promotion of rational recreation in the nineteenth century had rarely been driven by central government and was instead promoted by social reformers and municipal authorities. Furthermore, historians of leisure have shown that the organisation of recreational provision by non-governmental organisations has often operated as an informal extension of the state in Britain, thus circumventing the need for formal state involvement.[10] The NPFA certainly seems to fit with this conclusion. Its Council included two nominees of central government, but more significantly its governance and leadership structures included a long list of aristocratic elites, government ministers, cross-party representation of MPs and local politicians from across the country.[11] In addition, there was considerable continuity in the values and rhetoric used by earlier open space campaigners and the NPFA. The Metropolitan Public Gardens Association (1882) and Commons Preservation Society (1865) were both represented on the NPFA Council. William Melland, Manchester councillor and secretary of the Manchester and Salford Playing Fields Society, took on a key role in the NPFA, particularly in relation to children's playgrounds and play leadership.

The environmental campaigner Lawrence Chubb became general secretary of the NPFA in 1928, having previously been a prominent member of the Coal Abatement Society and the National Trust.[12] By 1929, both the NPFA and Commons Preservation Society operated from the same offices at 71 Eccleston Square in London.[13]

The NPFA initially made significant progress in the capital. Its first annual report recorded voluntary donations of £4.3m (£23,000) and successful negotiations with the London Underground for reduced fares to playing fields.[14] In Sutton and East Ham, it sought to ensure that play provision was included in the layout of new estates. Beyond London, the Birmingham Playing Fields Association helped to create a playground at Keeley Street, channelling £868,000 (£4,500) to help secure land in 'one of the most congested areas of the city'.[15] Schemes were initiated in other urban districts including Accrington, Wigan and Rochester. Although many of these schemes fell under the umbrella of the national association, there was nonetheless a strong sense of municipal civic pride, local voluntary action and small-scale philanthropy involved in making them happen. As such, the example of the NPFA and its local projects lend weight to suggestions that the urban remained a significant driver of citizenship in the early twentieth century and had not been entirely replaced by the national as is often assumed.[16]

Although the NPFA made much of the shortage of play space in densely populated urban areas, the countryside also featured prominently in its rhetoric and work from the outset. In its first year of operation, the NPFA were given land and money by private donors to create thirteen playing fields, of which five were in rural communities.[17] By the end of its second year, the county branch in Cornwall alone had been involved in ten schemes to provide village play spaces. And while green space campaigners had long drawn upon romantic visions of pastoral landscapes to inform the design and use of urban public parks, for the NPFA it was the village green that acted as a space of social cohesion and physical health. This early emphasis on the recreational needs of rural districts would be a consistent feature of NPFA campaigning in the interwar period, contributing to wider processes which saw rural communities negotiating the impacts of modern life.

At the same time, the NPFA also emphasised its role as a movement of modern times, advocating for the creation of well-planned, properly designed and technologically modern spaces for rural play and recreation. It sought to increase its sense of authority through nationwide coverage and by providing expertise and guidance, initially through its layout committee and later through *Playing Fields* and other design-focused publications. From the outset, the NPFA also made use of modern technology to promote its cause, particularly through regular national and regional radio broadcasts. The inaugural meeting of the NPFA was relayed from the Royal Albert Hall on BBC national radio.[18] Playing field associations from London and Glasgow to Monmouthshire and Gloucestershire regularly made appeals for funds on *The Week's Good Cause* radio programme, while the official opening of a playground was sometimes recorded on film.[19]

A 1934 article in *Playing Fields* provides a useful insight into the way that these seemingly contradictory values of tradition and modernity were played out in practice. The article explored the need for traditional village play spaces, but at the same time did not lament the loss of large country estates and associated ways of life. Instead, it focused on the consequences for play and recreation as new landowners no longer permitted the use of a meadow or field for villagers to play games. As estates were sold off and broken up, as happened at Eastchurch on the Isle of Sheppey in Kent, the children of the village were apparently left with 'absolutely nowhere but the roads on which to play'.[20] The unnamed author did not call for the retention of the country estate or oppose its subsequent redevelopment, but rather suggested that proper planning would enable land to be purchased at agricultural land values so that access to recreational facilities could continue. The implicit suggestion was that the idyllic vision of village cricket could be sustained through efficient planning and proper organisation. At the same time, it reiterated the well-established rhetoric that the street was a place of danger and the playground a place of safety. The landscape architect Marjory Allen, who features significantly in later chapters, also felt the village green represented an ideal place to play. Writing in 1937, she emphasised that where the village green had been lost to development, a sensitively designed playground could provide an entirely appropriate space for children's play.[21]

The most significant feature of the NPFA's attempt to develop a nationwide campaign was a tendency to identify the same problems and solutions in villages, towns and cities. The threat to existing open spaces and the dangers of the street were positioned as problems facing both urban and rural communities, and dedicated places for play were identified as the solution in both places too. But while it was possible to create playing fields on the edge of towns and in the expanding metropolitan suburbs, an alternative type of space was needed for young children living in the central areas of cities, where age or adversity limited access to distant sports pitches. The children's playground close to home was a pragmatic response to the processes of urbanisation and diminishing access to local common land. However, while the NPFA's guidance and model designs varied according to the size of the area available for play, they seldom varied according to its surroundings. Consequently, an inner-city local authority intending to create a playground received much the same advice as a rural parish council. Detailed design guidance invariably focused on the provision of manufactured playground equipment, such as swings and slides, and might also suggest a sandpit or a paddling pool.[22] By not differentiating between urban and rural play spaces and promoting the inclusion of manufactured equipment, the NPFA contributed to the standardisation of play space across Britain. This version of the playground would come to dominate both professional and public expectations of children's play spaces for at least the next fifty years.

Within this tendency towards standardisation and the more efficient provision of playgrounds was a complicated gender dynamic. On the one hand, gender-segregated play spaces had largely disappeared by the 1930s. The need for separate playgrounds for girls and boys had been a regular feature of late nineteenth-century rhetoric and the existence of such segregation is evident in park regulations and photographs. By the 1930s, only one of the LCC's forty-nine playgrounds included separate play facilities for girls and boys, while the Wicksteed Park playground and many others were also not segregated by gender.[23] The rhetoric used by NPFA campaigners seemed to treat girls and boys equally too. At the inaugural meeting, many of those who addressed the gathering spoke of the need to provide facilities and opportunities that would enable all children to participate.

At the same time, campaigners continued to emphasise prewar gender norms in their work to promote play spaces. In both conceptual and practical terms, the NPFA's approach was highly gendered and inequitable. The NPFA's fundamental assumption that leisure was the binary opposite of work or school failed to recognise the complexities of lived experienced for many older girls and women. For some, the park or playground may have provided a legitimate way to escape from the confines of home, while time spent with their children could be a source of pleasure for mothers. But, as Langhamer argues, 'child-centred forms of leisure, such as ... a visit to the park, should be viewed as a complex synthesis of both duty and pleasure for adult women'.[24] Meanwhile, in practical terms, the potential for older girls and young women to play was considerably undermined by the NPFA's emphasis on facilities for team sports, and in particular football. In 1921 the Football Association effectively banned women's football, despite a successful thirty-year history, and in doing so did much to establish a long-running social taboo surrounding women's participation in sport.[25]

The NPFA did little to challenge this taboo and instead focused on providing facilities for male-dominated sports. Dominant social norms most likely shaped this approach, but so too did the preponderance of public school- and military-educated men among the organisation's officers and committee members. In 1934, the patron of the NPFA was the king, its president the duke of York, and its officers included three earls, a field marshal, an admiral and two MPs, who had all been educated in public or military schools. Team games had been central to the culture of public schools and the military from the mid-nineteenth century, helping to explain the particular importance attached to sport among elites educated in these institutions.[26] Participation in games had started as a tool for managing pupil behaviour, but was soon understood as vital for the development of appropriately masculine character traits in schoolboys and cadets who would grow up to operate and administer the nation and empire.[27]

These idealised character traits were often embodied in the figure of the imperial soldier hero, a character often assumed by scholars to have disappeared as an ideal type during the First World War.[28] However, there is increasing doubt among cultural historians over whether this is indeed the case.[29] An examination of the rhetoric used by the NPFA certainly supports this revisionist view. Much like the contemporary boys' club movement, the NPFA continued to promote prewar notions of masculinity well into the 1930s.[30] This idealisation of a muscular, duty-bound, stoic and adventurous masculinity was often accompanied by a lampooning of the suburban, domesticated man and the NPFA was explicitly disparaging throughout the interwar period of this apparently feminised male character. These notions of gender were also combined with conceptions of class, and working-class masculinity in particular was characterised as deficient. The most notable examples are from the illustrations used on the cover of the NPFA's journal, where the physical degeneration associated with urban working-class life is visibly contrasted with the fitness and stature of a heroic, middle-class sportsman. The front cover of *Playing Fields Journal* from 1934 showed a stooped working man, coming from the polluted air of the city, being welcomed by an upright, muscular footballer to the clear skies and tree-lined playing field. Much like the rhetoric of the MPGA in the 1890s, the NPFA continued to associate the problematic urban environment with notions of working-class degeneration, while also drawing on conservative gender ideals.

Furthermore, just as late nineteenth-century campaigners had been partially motivated by the apparent inadequacy of potential military recruits, so too were open space campaigners in the interwar years. Despite reservations about the quality of medical statistics, the idea that wartime conscription was constrained by the poor physical health of potential recruits remained a powerful rhetoric into the 1930s among playground advocates.[31] Both the military classification system (where those classed A1 were fit for overseas service, B1 for garrison duties, C3 for sedentary duties and so on) and the association between stature and physical fitness remained powerful markers of health. Writing in 1935, Edward Prentice Mawson felt that Britain could never again 'be caught with a predominantly C3 population' and that better designed public play spaces would help to address this shortcoming.[32] However, beyond play space advocates, these anxieties about physical fitness resulted in a focus on the bodily health of adults rather than children, even as adult health became increasingly associated with youth-preserving exercise.[33] Within the evolution of a wider physical culture movement, the 1937 Physical Training and Recreation Act provided grants to develop recreational amenities for adults, but specifically excluded facilities for children.[34] Instead, the somewhat ambiguous connections between physical health, class, gender and the urban environment would be replaced in the rhetoric of play space campaigners by a more direct and obvious threat to children's lives.

Safety and supervision

While earlier notions of masculinity and assumptions about city life endured during the interwar period, there was also a gradual change in the way that the NPFA perceived and explained the threats that children faced in the modern world. The somewhat hazy connection between city life and individual physical stature was increasingly superseded by the use of compelling statistics which revealed the direct threat to children's lives from the increased number of motor vehicles on both urban and rural roads. Contemporary responses to the dangers of the street were diverse, but the increasing dominance of motor vehicles was rarely questioned. Instead, responsibility was implicitly placed on children to adapt their behaviour to this changing public environment. Road safety training, the creation of play streets and even arrests were all part of this wider response, but the NPFA emphasised that the best solution was to remove children from the streets altogether.

Street play was still a frequent activity for children in the interwar period. The playgrounds which had been created by this point could not meet the recreational needs of all neighbourhoods, with dedicated play spaces too distant and other open spaces too formal. From personal accounts of growing up in London we know that the streets outside children's homes invariably offered opportunities for sociability, play, spectacle, financial reward, and for older girls in particular, the responsibility of childcare.[35] On a brief summer stroll in 1928, the Bishop of Southwark counted 'twelve games of cricket, six games of rounders, several mysterious games which consisted of hopping from square to square' all taking place in the streets of Kennington.[36] The use of the street as a de facto playground had long been a feature of urban life, but what changed in this period was a data set of newly available statistics that highlighted to campaigners a stark indication of the dangers of the street.

In 1919, there were around 300,000 motor vehicles using the roads, a figure that increased to over 3 million by 1939.[37] This increase in the number of motorised vehicles also resulted in an increasing number of collisions with children. In 1931, *Playing Fields* reported on the work of the London and Home Counties Traffic Advisory Committee and its report 'Street Accidents to Children in Greater London'. It found that playing in the street was the second most prolific source of motor accidents and that children between the ages of five and nine were most likely to be the victims of collisions.[38] A year later, the NPFA joined a deputation to the Minister of Transport to protest at the 6,000 fatalities that occurred on the roads in 1931, including many child victims.[39] By October 1932, *Playing Fields* was emotively describing 'the cry of the children', as hundreds of child deaths and 10,000 injuries on the roads each year resulted in 'a pitiful tragedy of family bereavement or crippled life'.[40]

The 1936 report of the Interdepartmental Committee on Road Safety Among School Children refined the data and statistics even further. It found that child road deaths had increased from 857 in 1920 to 1,433 in 1930, while the percentage of child fatalities which occurred on the road had increased from seven per cent in 1903 to forty per cent in 1933.[41] *Playing Fields* followed this up with a piece which showed how the data could support the play space cause. In an article which promoted 'the case for playing fields from a new angle', LCC education officer Mr Lowndes emphasised the financial costs of child injuries and deaths on the road and the spatial relationships that could help explain them.[42] At a time when many of the consequences of traffic accidents, such as hospital treatment or an early pension due to ill health, were not paid for from public funds, this was a call for playgrounds as a more efficient way to organise society, rather than necessarily a way to save public money specifically. At the same time, statistics from the Interdepartmental Committee showed that a case could be made for a close correlation between access to open space and child casualties; London Boroughs that were covered by more open space had a lower proportion of child road victims. The creation of new playgrounds assumed a renewed importance in this context: not only healthy and character building but lifesaving too. For Dr Mabel Jane Reaney, child psychologist and active member of the NPFA, it was

> often the adventure-loving child with initiative and enterprise who is the victim, so that the nation is deprived of another potential leader. Scarcely a day passes without a coroner pointing out that the life of a child might have been spared if it had not been playing in the street.[43]

The Interdepartmental Committee concurred with this view, stressing the need to balance the protection of children from harm while not inhibiting the 'spirit of adventure' that was both inherent in many children and a valuable national characteristic.

While there was growing concern about the rising number of child road deaths and the consequences for both families and the nation, there was also growing acceptance of the role of motor vehicles in modern society. Child behaviour and not the internal combustion engine, it seemed, was the essence of the problem. Accordingly, during the 1920s and 1930s both public and press opinion gradually shifted from generally siding with pedestrians to seeing them as increasingly unpredictable and erratic.[44] In an inquest into the death of three children under the age of three who had been run over by motor vehicles, the coroner focused on how 'it was not fair to drivers that parents should allow their children to play in the street'.[45] In 1928, a newspaper columnist emphasised the 'unbearable strain' placed on motorists by children's street play, emphatically stating that 'there is no factor which plays so devastating a part in the wrecking of a motorist's nerves as does the heedless child'.[46] Attempts were made to manage the use of roads and streets by motor vehicles. For example, the 1934 Road Traffic Act reintroduced a 30mph speed limit in built-up areas, but on the whole, it was children and their behaviour that was problematised. As a result, attempts to solve the problem of 'traffic accidents' were invariably focused on marshalling children and their play, rather than challenging motorists' use of public space. Most significantly here, the children's playground came to feature in several, although not all, responses to the problem.

The most uncompromising response to the problem of child road deaths was to forcibly prevent children from playing in the street. Section 72 of the 1835 Highways Act had long made it an offence to play on a public highway. In the interwar period, this section of the Act was still used to discourage children's street play and, in some cases, to remove children from the street altogether. Admittedly, the usual police practice, in London at least, was for an inspector to caution the child in front of their parents, apparently with 'good moral effect' on the child and great appreciation from the adults.[47] Even so, of 1,828 cases that were heard in the eight Metropolitan Juvenile Courts in 1930, over a third related to playing in the street, while in the subsequent five years over 1,000 cases relating to street play were heard in court.[48] This practice was not without its critics. For Nancy Astor MP, speaking in the House of Commons, there was 'no more pitiable sight in life than a child which has been arrested for playing in the street. Of all the pitiable sights that I have seen, that is the most pitiable. Though these children may be fined, we stand convicted.'[49]

A second response involved attempting to educate children to cope with life on the streets, by bringing the road into the playground. From the 1930s the British government promoted road safety education through school crossing patrols, children's clubs and public education films. At a local scale, parents also campaigned to improve road safety in their neighbourhoods. In north London, the Seven Sisters Safety Committee fought for safety improvements in Tottenham following the death of two five-year-old girls within a fortnight, both killed on the road by lorries.[50] A year later, a repurposed version of the playground became a novel attempt to educate the children of north London in road safety. The first 'model traffic playground' was opened in Lordship Lane Park in Tottenham in 1938. Designed by G.E. Paris, the borough's parks superintendent, it was officially opened by the Minister of Transport and received widespread media publicity.[51] As a training ground for an automotive society, the traffic playground included nearly a mile of roadway and miniature highway features, including traffic lights, police callbox, road signs and pedestrian crossings, so that it would resemble the conditions children could meet on real roads (Figure 3.1). Children were able to hire model cars or bring their own bicycles to use on the roadways, while playground equipment was located so that other children had to cross the road to get to it.[52] While the traffic playground was ostensibly designed to educate children in road safety, Superintendent Paris later acknowledged that it was also in part a response to the problem of children annoying adults by riding their bikes around parks.[53] The model traffic playground was hugely popular with local children and was open until the outbreak of the Second World War, when it closed for nearly ten years.[54] Several other traffic playgrounds were created after the war, including in Dundee, Salford and Scunthorpe, but the adoption of the playground as a tool in road safety education did not become widespread.[55]

A third response to the problem of child road deaths reversed the assumptions that underpinned the model traffic area and instead brought the playground into the road. Although popular opinion may have been

Figure 3.1: Children's traffic playground, Tottenham, 1938, © Daily Herald Archive / Science Museum Group.

increasingly sympathetic to motorists, a small minority felt that expecting children to take responsibility for their safety on the streets was both unreasonable and unlikely, challenging the prevailing attitude that 'if children are killed, it is their own fault'.[56] Instead, a more radical response involved excluding motor traffic from the streets where children played. The earliest attempt to create safer streets for children's play in Britain has been associated with Salford's Police Chief Constable, Major C.V. Godfrey. He too felt it was impossible to train younger children to keep themselves safe on the roads and the only rational solution was to prevent motor vehicles from using streets when children were most likely to be playing out. As a result, by 1929 over one hundred streets in Salford were closed to through traffic after school.[57] During the 1930s the idea gathered momentum. In 1930, the London and Greater London Playing Fields Association contacted the Metropolitan Police to explore the possibilities of emulating a successful New York street play scheme in the congested areas of London.[58] City authorities in New York had first experimented with closing Eldridge Street to motor traffic to create a space for play in 1914 and by 1929 one hundred and sixty-five play streets had been created in thirty-six cities across the USA.[59] By 1934, the London Society and the London Safety First Council were calling for street playgrounds to receive greater consideration in the capital.[60] In 1936, the Interdepartmental Committee on Road Safety Among School Children recommended that legislation was needed to enable local authorities to create play streets, particularly for congested neighbourhoods without access to adequate play spaces, eventually resulting in the Street Playgrounds Act of 1938.[61] By 1950, seventeen local authorities had closed streets for play, with a further eight closures under consideration by the Minister of Transport.[62]

For the NPFA, street playgrounds were never an ideal solution to the problems of urban childhood. Playing on the street was seen as unhygienic and a threat to nearby property, while officially sanctioning such activities could even disincentivise local authorities from providing 'proper' play spaces. Moreover, roads represented the economic and circulatory drivers of city prosperity and were increasingly accepted as adult, automotive spaces. For a while at least, the problem of providing play space in congested inner cities outweighed these objections and the NPFA lobbied privately in favour of play streets. In 1932, the NPFA produced a draft Private Members Bill for discussion in parliament and pressured the Home Office, Police and Ministry of Transport to act.[63] But by the mid-1930s, the NPFA felt compelled to publicly distance itself from the play street campaign. An editorial in *Playing Fields* in January 1935 firmly stated that closing streets for play was both contrary to organisational policy and inappropriate: 'it is obvious that it is not desirable to create in the mind of any child the impression that a street is a natural or proper place for play'.[64] In this instance, it seems that the NPFA eschewed pragmatism in favour of an apparently 'natural' principle. But while this self-belief and call to natural principles may have limited the uptake of play streets in a few instances, it also contributed to the NPFA's subsequent commitment to the creation of 'proper' places to play.

For the NPFA, the only proper response to the problem of child road deaths was to create dedicated spaces that would encourage children to play away from the dangers of the street. Furthermore, such spaces needed to adhere to campaigners' expectations of an appropriate space for play, increasingly a vision dominated by playground technology. As such, it was the presence of manufactured playground equipment, appropriately arranged and safely installed, that determined whether spaces would fulfil their function as a site of safety. This approach chimed with wider expectations about children's place in society and generated significant support for the NPFA and its cause. Only a few years after its official inauguration, the Association secured considerable financial resources to help deliver its vision and increasingly assumed a position of authority in the field of children's play. Within a year of its first national appeal in June 1927, the NPFA received over £63m (£330,000) and gifts of 157 hectares (388 acres) of land.[65] The most substantial financial support came from the Carnegie UK Trust, which contributed £38.6m (£200,000) as part of its wider efforts to 'give a lead to important new movements of a national character ... which at any given moment appear to be of prime importance'.[66] The Trust was established by the Scottish-American industrialist Andrew Carnegie in 1913 and had previously supported the creation of public libraries, child welfare schemes and other educational and cultural activities. With widespread political support for the appeal, earlier donations from the king and queen, and the close match between the objectives of the Carnegie Trust and the perceived value of playgrounds, the donation was understandable, even if the cash amount was substantial.

The Carnegie donation was to be distributed as one-off grants that could only form part of the funding for any given scheme, helping to establish a role for the NPFA in allocating financial support that would continue until the 1960s.[67] Because Carnegie and the NPFA only part-funded schemes, local authorities also needed to raise money from elsewhere, often in the form of loans from central government. One estimate suggests that between 1920 and 1935, the Ministry of Health sanctioned loans for the purchase and construction of parks and playgrounds worth over £4.4bn (£22m).[68] Following the success of its national appeals and its administrative role in distributing grants, the chair of the NPFA, Sir Noel Curtis-Bennett, was able to assert that it had supported the creation of 782 children's playgrounds and over 1,000 other recreational facilities in the interwar years.[69]

As well as its involvement in financing play spaces, the NPFA issued advice and guidance on the design and layout of playgrounds to ensure they attracted children away from the street and fulfilled their wider functions. From the outset, the NPFA had established a layout subcommittee to provide expert advice and guidance to local authorities and others interested in play space design.[70] In addition, most interwar issues of *Playing Field* included articles or commentary specifically on children's playgrounds, along with advice on their arrangement, equipment and supervision. Advice ranged widely in scope, from advocating 'the demolition of one or two small houses' to provide playgrounds in congested areas, to the repurposing of churchyards as spaces for play, and generally reinforced the view that playgrounds were ideally situated in 'some odd corner or other' to keep children out of the way.[71]

In a 1930 article, Brigadier General Maud, chair of the layout committee and chief officer of the LCC parks department, argued for the greater provision of equipped playgrounds, specifically in public parks. He felt that it was not enough to simply provide grassed areas for children to play and instead argued that

> the best way to keep them out of mischief and amused is to provide gymnasia ... well equipped with swings, giant strides and the various forms of modern apparatus now on the market, not forgetting a sand pit ... tucked away so that those who do not seek it out will be unconscious of its existence.[72]

At the same time, *Playing Fields* included an increasing number of adverts for play equipment manufacturers from across the UK. This combination of commentary, guidance and advertising meant that the NPFA both contributed to and reflected changing ideas about the way that play spaces were imagined.

Manchester parks superintendent W.W. Pettigrew asserted that 'heavy iron chain swings' had long been the most popular piece of playground equipment and until 1914 were 'practically the first equipment provided in every town playground as soon as it was acquired'.[73] From the 1920s, the increasing number of suppliers and diversity of products meant that playgrounds included a wider range of equipment, although the swing had become a key feature in the signification of dedicated places for children to play, and in some cases dominated entire open spaces. LCC parks department records from the late 1930s show that of the ninety open spaces they managed with recreational facilities, forty-nine had dedicated spaces for children to play. These playgrounds included a wide range of manufactured equipment, from giant strides, see-saws and rocking horses to merry-go-rounds, maypoles and ocean waves. Forty-three included a sandpit and twenty-six had a paddling pool. But the sheer number of swings is the most remarkable feature of many of these play spaces. Most included an individual pendulum swing or plank swing, but regular swings were still by far the most numerous, with over 1,000 provided in total. Large parks, such as Victoria Park (87ha) or Southwark Park (25ha), had over fifty swings each, but even smaller spaces were abundantly equipped. The one-hectare Newington Recreation Ground in Southwark included forty-five swings, leaving little room for other playful activity.[74]

Aside from this proliferation of swings and play equipment, the question arose as to who was best placed to exercise leadership over these newly engineered play spaces. As we have already seen, the MPGA had argued for caretakers who might prevent bad behaviour in children's playgrounds. By the 1920s, this approach was being challenged by ideas from the developing fields of child psychology and progressive education, which emphasised a supportive rather than supervisory role for adults.

However, this challenge was far from homogeneous, particularly among progressive educationalists, and there was considerable debate about the extent to which adults should intervene in children's playful activities. Indeed, the activities of the NPFA in these years highlight the extent to which campaigners sought to rework earlier notions of the playground as a space of physical exercise, adult supervision and order, on the basis of emerging ideas which emphasised play experiences, adult facilitation and varying notions of freedom.

The interwar period saw a significant growth in psychologically informed approaches to caring for children, including child guidance clinics, progressive nursery education and play groups. Although rarely explicitly acknowledged by play space campaigners, the evolution of ideas about child development and the emerging field of psychology helped to shape a more nuanced understanding of children at play, even if the impact on play spaces was less immediately discernible. As early as 1908, the Board of Education Consultative Committee recognised that children under five needed specific educational spaces which encouraged movement, play, variety and rest, ideas that were implemented in practice by the nursery school movement.[75] The assumption that children had specific requirements from education and play at different stages of their lives remained central to the work of advocates such as Susan Isaacs, the hugely influential educational psychologist, and the socialist journalist and nursery education campaigner Margaret McMillan. Neither Isaacs nor McMillan seem to have been directly involved in the NPFA or wider attempts to promote public play spaces. But play and nature were important themes in their work, and it seems likely that they had some influence on ideas about how and where children should play.

McMillan is best known as a leading campaigner for the provision of nursery schools, but she was also an active member of the Independent Labour Party and Bradford School Board, a socialist journalist and founder of the Deptford Clinic in south London. While working in Bradford, McMillan introduced child-size furniture and sand trays into school classrooms. After moving to London, she shared social circles with members of the university settlement movement, including Philip Wicksteed, Charles Wicksteed's brother. In 1911 she established the Deptford Clinic, where children were treated for a variety of conditions and initially slept outside in the garden of the clinic to recuperate.[76] For McMillan, the clinic garden soon became much more than simply somewhere to recover, providing clean air, brightness and movement, as well as protection from the corrupting influence of society and the dangers of nature. For a time, the garden included apparatus to climb on, a lawn to run about on, a sandpit and a rubbish heap, with stones, old iron, assorted pots, and no rules, where children could play freely.[77] During the 1920s the role of the garden diminished both in the work of the clinic and in significance for McMillan, but despite this it seems to have mirrored evolving ideas about the function and design of public play spaces for children, especially the combination of health and education, physical apparatus and sand play. In contrast, playing with junk, in spaces such as adventure playgrounds, would not feature more prominently in either the ideas or practices of play space campaigners until after the Second World War.

The influence of interwar educational psychologists on playgrounds was in some ways less obvious, but nonetheless significant. By the early twentieth century, educational psychology was a customary feature of most teacher training courses. With foundations in the theories and practices of Freud, Froebel, Dewey and Montessori, individuals such as Susan Issacs and A.S. Neill were promoting and experimenting with alternative approaches to childhood health and education. Isaacs was particularly associated with the nursery movement and shared assumptions with McMillan about the role of infant schools as sites of education, health and social reform. Her practical experiment in progressive education, the Malting House School (1924–9), included a large garden, sandpit, tree house and tools for children to use.[78] A.S. Neill's Summerhill (1921) took child-centred education to its most extreme, with no formal lessons and an emphasis instead on unstructured play, voluntary participation in activities and an emphasis on individual freedom within a community setting.[79] For a time Neill taught at King Alfred's School in London, describing it as the freest school in England at the time, while subsequently arguing that 'the evils of civilisation are due to the fact that no child has ever had enough play'.[80] Joseph Hartley Wicksteed, Charles Wicksteed's nephew and later headmaster at King Alfred's, combined this focus on play and self-determination with an emphasis on the outdoors, suggesting that the 'garden, field, or woodland' were the ideal places for children's education.[81] These shifts in educational thinking and experiments in progressive education signalled the beginnings of a move away from rote learning, formal teaching and restraint, towards freedom, discovery and a more permissive approach to discipline.

But while these values may have had little direct influence on the form of playgrounds, they did shape wider attitudes to childhood, particularly an increased awareness of the emotional lives of children and the role of play in child development. The New Education Fellowship promoted these progressive notions of childhood, while the wide-ranging careers of advocates, such as Isaacs and psychologist Cyril Burt, also helped to popularise aspects of progressive education.[82] Their work in academia, education and popular communication meant they had a 'powerful and enduring influence' on notions of childhood.[83] The psychologist, writer and active NPFA member, Mabel Jane Reaney, also promoted these values more broadly, emphasising the place of play in the public realm. In 1919 she argued that there should be a director of play within government to ensure that towns and cities provided spaces for children to partake in 'free play'.[84] In her 1927 publication, *The Place of Play in Education*, Reaney stressed the significance of free, 'natural' play in children's mental and physical development, while her later articles and conference presentations promoted these values to both popular and specialist audiences.[85]

For playground campaigners, the arena where these emerging ideas had the greatest influence was in discussions about play leadership, an issue that was particularly problematic for the NPFA. Adult involvement in the use of playing fields was straightforward – the rules, regulations and norms of adult team games were well established and could easily be applied to the games of older children and young people, with adults as referees. In contrast, the rules of the playground and the way that adults could or should be involved in children's informal play had not yet been settled. In an attempt to make sense of the issues and to recommend a way forward, the NPFA established a subcommittee on play leadership in December 1928.[86] The committee was chaired by William Melland, Manchester Councillor, member of the city's parks committee and leading figure in the Manchester and Salford Playing Fields Society. Other progressive members of the subcommittee included Miss Spafford from the Ling Association, which had promoted Swedish-inspired gymnastics and physical education since 1899, and Mabel Jane Reaney. The committee also included the apparently authoritarian and insensitive Commander B.T. Coote, Welfare Advisor to the Miners' Welfare Fund (an endowment created from colliery company subscriptions to provide welfare facilities for mining communities including, for example, children's playgrounds in the village of Llanbradach in Glamorgan and Newtongrange in Midlothian).[87]

From the outset, the play leadership committee worked to emphasise the urgent need for a systematic approach, the type of people who would make appropriate leaders, the training they needed and how it could be organised, as well as the need for coordination with both the Board of Education and Local Education Authorities. In many ways the committee was clearer about the actions that were needed to develop and promote a system of play leadership than they were about fundamental questions relating to adults' role in children's play or wider uncertainties about the relative importance of nature and nurture, environment and society in children's development.

In July 1933, the committee organised a conference on play leadership to explore these issues, with over one hundred delegates from across the country attending. In his conference address, William Melland attempted to bridge the gap between notions of supervision and leadership. From his experience as a member of the parks committee, he felt able to distinguish between the day-to-day management of play spaces and play leadership; park keepers could ensure law and order were maintained but specialist staff were needed for the latter. Fellow delegate, Miss Spafford, went further in emphasising that adult guidance would ensure that the full benefits of the playground were realised. She had initiated the debate about play leadership within the NPFA, sending a memorandum on play leadership to the NPFA executive committee that prompted the creation of the play leadership subcommittee. She also presented at the conference and argued that a play leader was far more important than apparatus, echoing assertions coming from the USA two decades earlier that organisation rather than equipment was the most important feature of a successful playground. She explicitly emphasised the need to learn from the approach adopted in the USA where, by 1925, 17,000 paid play leaders were working in 8,608 play areas.[88] Her presentation also revealed the complexities around adult involvement in children's play. She felt that children's play should be 'free' and 'guided' rather than over-organised, but at the same time expected a play leader to 'prevent roughness and noise and make for order and discipline' through highly structured activities such as basketmaking and folk dancing.[89]

Mabel Jane Reaney also addressed the conference and warned against over-enthusiasm on the part of adults in organising children's games. Her conference presentation on 'the urgent need for trained play leaders' combined concern about the impacts of the modern world on children and the nation, highlighted the problems with existing playgrounds and proposed solutions inspired by the latest ideas in psychology and child development. In her critique of existing play provision, Reaney felt that playgrounds would be unused or mistreated unless they were properly supervised and organised. She argued that a more scientific understanding of biology, psychology and child development should inform play provision and the work of play leaders. She noted that children of different ages engaged in different types of play, an argument rooted in the idea that children gradually developed instinctive play tendencies. Seemingly based on the work of earlier child psychologists such as G. Stanley Hall, she categorised children's play into distinct 'play periods' and identified the activities and leadership that were necessary for each. Children from birth to seven years of age played individually and sand provided a good medium for their expressive play. From seven to nine years old, they became more active but still played individually. From the age of nine to twelve, children became more cooperative, and it was here that the play leader could make all the difference between 'a quarrelling herd and a self-respecting team'. Despite her claim that these traits were an innate part of every child, she also argued that play leaders needed to 'know the needs of the children and how to satisfy these needs – for the children do not know themselves'.[90]

Overall, the play leadership subcommittee argued that the instinct to play was biologically inherent in children, but at the same time found that children did not have the knowledge about how best to play, echoing earlier suggestions that 'we have to teach a nation unused and unapt to play'.[91] As a result, play leaders were necessary to ensure that children's play achieved everything that campaigners hoped it could. The work of the committee was given greater emphasis as the economic crisis of the early 1930s led the NPFA to cut staff and reduce spending. While funding was not available for creating new playgrounds, the NPFA could promote the importance of existing facilities and emphasise the role that play leaders could have in maximising their usefulness. At the same time, local authorities faced even harsher economic conditions and could rarely justify recruiting additional staff, prompting a debate within the NPFA about the merits and shortcomings of volunteer play leaders.[92] By 1936, the committee reported that six play leadership courses were taking place, while a London course had just finished with over sixty students graduating.[93] Demonstrations in play leadership continued, including events in Leamington, Leicester and Folkestone, the latter in conjunction with the annual conference of the Institute of Park Administration. As these locations suggest, play leadership remained largely an urban phenomenon. Organisations such as the Kent Rural Community Council did run lectures in play leadership, but more comprehensive training schemes were mainly confined to larger cities, including Birmingham, Glasgow and London.[94] Although ideas about play leadership did not achieve the same prominence as the principle of providing play spaces, the interwar meeting of progressive educationalists and playground campaigners set the conceptual foundations for the mid-century adventure playground movement and the play worker profession. Just as debates were taking place about the role of adults in children's play, the physical form of the playground and the notion that it was a healthy and safe space were being challenged too.

Problems in the playground

As earlier chapters have shown, the creation of playgrounds could be controversial and invariably echoed wider debates over the use of public space. In the interwar period too, playgrounds were not always a welcome addition to green spaces and were sometimes perceived as a threat to traditional visions of the way that urban parks should be designed and used. In late 1929, for instance, the NPFA became embroiled in a public dispute between park traditionalists and George Lansbury, an east London MP and recently appointed First Commissioner of Works in the second Labour government. As Commissioner, Lansbury was responsible for the crown estate and soon 'amazed civil servants with a radical programme of recreational improvements for the public in the royal parks'.[95] His programme, styled Brighter London Parks, focused on providing sporting and cultural facilities and dedicated play spaces for children. By October 1929 he was able to declare that 'work would be begun almost at once on providing sandpits, swings, ponds, and shelters'.[96] The NPFA were keen to support the initiative and offered over £960,000 (£5,000) to help convert the former exhibition grounds in Hyde Park into playing fields, augmenting three existing sports pitches used by soldiers from the adjacent Kensington Barracks with a further three, and highlighting the enduring connections between representations of sport and military conflict.[97]

In response to the plans, an editorial in *The Times* expressed in thinly veiled class terms unease about 'ill-considered innovations' and 'mysterious operations' that would convert 'one of the most beautiful green stretches of Hyde Park' into a 'monstrous imitation of Coney Island or the beach at Blackpool'.[98] Lansbury's proposals challenged traditional assumptions that emphasised the place of wholesome and genteel recreation within the park boundary, rather than playful amusement. The plans were debated in the House of Commons and while some felt Lansbury was humanising the office of the First Commissioner of Works, others were concerned about his proposals and their impact on park aesthetics.[99] Anxious correspondents wrote to *The Times* of their horror at the intrusion of football goals and the threat that the wider proposals posed to the principles of 'peace and beauty and freedom' which they felt defined the parks.[100] Lansbury felt strongly that a wider public needed facilities in the royal parks and that he 'could not desecrate them by having too much for the children'.[101] NPFA supporters endorsed his plans both in the press and in parliament, stressing that 'the poor boys and girls of dismal Camden Town' deserved places to play close to their homes and that a paddling pool, sandpit and see-saw would take up very little room.[102]

The playground had become embroiled in fraught negotiations about the fundamental principles and purposes of public parks and the place of children in such spaces. It seems that Lansbury was committed to providing better public facilities for Londoners, but the programme of works also provided an opportunity for him to challenge traditional, elitist visions of the royal parks as bucolic, socially exclusive, natural landscapes in the heart of the city. Both Lansbury and the NPFA emphasised that children had a legitimate claim on public space, contributing to ongoing conversations about parks for the people and what that means. They implicitly argued that parks and playgrounds were about making the city more liveable for everyone and that the provision of dedicated places to play represented an important way for society to recognise and facilitate that claim.

While the introduction of playgrounds may have challenged traditional ideas about what public parks were for and how they should be used, once installed and in use they could also challenge campaigners' rhetoric too. In principle, the playground promised health, education and safety for children, but in practice it could be a space of accidents, incidents and potentially a vector of ill health. Playground campaigners invariably emphasised a simple binary between dangerous streets and safe playgrounds, but in reality, dedicated play spaces had long represented spaces of risk to children. Early twentieth-century newspaper reports show that a falling bell, an explosion and a runaway horse all injured and killed children in purportedly safe play spaces.[103] The public playground, too, could be a space where children were exposed to life-changing and sometimes fatal events.

Before the First World War, local newspapers occasionally reported on accidents in the playground, briefly noting court proceedings and victims' claims for compensation. For example, ten-year-old James Prosser sustained injuries after he fell from a swing in McLeod Street Playground, Edinburgh; a six-year-old girl was injured after being hit by a swing seat in Manchester's Queens Park; and Albert Davage, aged nine, was killed in Charlton Kings Playground in Gloucester when a giant stride collapsed.[104] The outcome of these cases was by no means certain as the legal system attempted to establish principles about who was liable for accidents, the parameters against which these cases could be judged and the damages that might reasonably be awarded.

Within a wider landscape of interwar concern for safety, the problem of playground accidents became more pressing. Heightened anxiety about the human and financial costs of industrial and road accidents resulted in the creation of a number of organisations that campaigned to improve safety, including the London 'Safety First' Council (1916) which in time became the National 'Safety First' Association and ultimately the Royal Society for the Prevention of Accidents (RoSPA).[105] Playground accidents in particular received greater publicity during this period, again in relation to the issue of legal liability and associated financial compensation for victims. It is difficult to conclude whether injuries and fatalities in playgrounds became more numerous, but concern and coverage both seem to have increased (the physical harm caused to children by such accidents did not generate wider concern, nor the interest of RoSPA and others, until the 1960s).

Play equipment had long balanced a fine line between providing exhilaration and ensuring safety. Manufacturers often emphasised the modern technological features of their products, including Wicksteed's 'hydraulic non-bumper see-saw' and 'patent safety arrangement' that prevented over-swinging on the Ocean Wave, as well as Spencer, Heath and George's 'safety coaster slide' with safety rails and wire cage underneath.[106] But despite these claims to technological advancement, children were still injured and sometimes even killed while playing in the playground. In 1931, the NPFA felt compelled to include a commentary in *Playing Fields* on the problem of legal liability for accidents and highlighted that the Miners' Welfare Committee had already secured insurance for its play spaces.[107] As a result, the NPFA worked with insurance brokers to organise a specialist policy that other open space managers could purchase to help mitigate the financial consequences of legal proceedings linked to playground injuries.

Despite this attempt to provide reassurance, reports of court cases where damages were awarded to injured children and their families continued to 'occasion a good deal of alarm' among park managers throughout the 1930s (and beyond).[108] Presumably this alarm was in part caused by the level of damages sometimes awarded by the courts when playground providers were found to be at fault. When four-year-old Peter Coates was paralysed after being injured on a slide in Rawtenstall Recreation Ground in Lancashire, a judge initially awarded £295,000 (£1,500) in damages, which was subsequently doubled on appeal.[109] Even in the face of sizeable awards of damages, some park managers felt that it was probably more cost effective to accept the risk of a claim than to go to the expense of modifying equipment or increasing the number of playground attendants.[110] In general, cases seem to have been judged against the principle that authorities were not liable for the consequences of dangers that were reasonably obvious to children.[111] But even though authorities were seldom found liable for playground accidents – and the NPFA keenly promoted such 'successes' – nervousness among park managers has been an enduring feature of public playground provision.[112]

Another threat to both individual children and the idea of the playground as a safe and healthy space was the 'indecent' behaviour of a small number of adults, resonating with enduring anxieties about the unhealthy aspects of public space use. As we saw in Chapter 2, the Metropolitan Radical Federation and LCC education committee had both previously raised concerns about such issues. But while the number of reported cases may have been low, for those children who were victims of sexual assault the issue was understandably distressing and often mentally and physically harmful. The written reports produced by park keepers provide troubling accounts of children's experiences, often recording perpetrators' attempts to entice children away from the playground, but also documenting the details of sexual assault.[113] The accounts record male perpetrators and female victims, contributing to a sense that male strangers presented the main threat to children. In Manchester, the parks department recognised the need to take action to prevent men from 'interfering with or molesting children' and so, somewhat clumsily, banned children's fathers from entering playgrounds with the rest of their family.[114] In London, the LCC education officer began coordinating the response of park authorities and the police to cases of sexual assault and also attempted to support victims in the aftermath of such crimes.[115] Although the number of cases appears to have been very low, their social significance ensured that the connection between the playground and fear of strangers would be a recurring one.

Just as dangerous adults and defective equipment could make the playground a hazardous place, so too could less-visible perils. These threats were often associated with the fabric of the playground and in particular the materials that formed the surfaces where children played. Gravel-surfaced areas for games, also known as dry playgrounds, were installed in a number of London parks in the early 1920s, including Clissold Park and London Fields. Soon concerns were raised by the Medical Officer of Health about children inadvertently inhaling dust while using them, leading the parks department to carry out a number of experiments in an attempt to tackle the problem.[116] Much more troubling than dust, however, was sand.

Although seaside resorts, particularly those with extensive sandy beaches, were often characterised as spaces of restorative health, when transported to the city the relationship between sand and health was far less straightforward, particularly for park managers.[117] The propensity for sandpits to harbour dirt, debris and disease-carrying pests meant that for some park staff they were seen as a 'menace to health'.[118] Inspired by anthropologist Mary Douglas's conception of dirt as matter out of place, Canadian geographers Ann Marie Murnaghan and Laura Shillington have recently argued that our conception of urban sand has shifted over the course of a century, from purposeful to problematic.[119] They have suggested that in the late nineteenth century sand had a rightful place in the city as a symbol of education and health, whereas by the late twentieth century urban sand was increasingly understood as unhealthy, dirty and out of place. But sand's journey from pure to pathogenic has not been a straightforward, gradual, linear transformation. Instead, from the outset public sandpits have been seen as troublesome spaces where insects and infestations could lurk, as well as hopeful sites of learning, interaction with nature and occasionally even entertainment.

The idea that sand could be educational had its origins in Friedrich Froebel's child-centred educational theories of the early nineteenth century. In his kindergarten, originally a metaphorical rather than a physical space, a shallow box of sand inside the classroom provided a practical tool for helping young children to develop physical skills.[120] Joseph Lee, the president of the Playground Association of America, would mystically claim that playing with sand connected children with their primeval amphibious ancestors, reflecting contemporary ideas about the evolutionary role of play in child development.[121] In Britain, the organisation that promoted Froebel's pedagogy, the Froebel Society, had endorsed the value of sand as an educational tool since its formation in 1874, but the first mention of sand in a public park seems to be in May 1893, when it was seen as something that was primarily enjoyable rather than necessarily instructive. The superintendent of Victoria Park reported to the LCC parks committee that 'during the year a novel mode of enjoyment has been provided for the little children of the East End of London in the form of a sea-sand pit which is apparently much appreciated by the little ones' (Figure 3.2).[122] This imagined connection between the urban sandpit and the seaside endured and seems to have reflected wider ideas about both the role of parks in bringing nature into the city and the recreational value of public spaces. In 1898, J.J. Sexby, parks superintendent of the LCC, remarked that

> nothing can be pleasanter than to stroll round from point to point and watch the happy little crowds disporting themselves on swings and seesaws, sailing their boats on the waters of the lake, of digging in the sand-pit, apparently quite as happy as though they were within sight and sound of the sea-waves.[123]

However, the image he shared of the sandpit in Victoria Park, with well-dressed adults and orderly rows of well-behaved children, suggests a

Figure 3.2: Children playing in sandpits, Victoria Park, London, 1893, © London Metropolitan Archives (City of London), SC/PHL/02/1141/B2895.

structured form of recreation that reflected traditional ideas about rational park behaviour.

As these values were gradually replaced in the playground by an emphasis on enjoyment and excitement in the early twentieth century, the sandpit might have become redundant. But while expectations around the value and use of sand shifted, sandpits became increasingly common in both the rhetoric of campaigners and play spaces themselves. Writing in 1907, the journalist Annesley Kenealy felt that a heap of sand was an essential feature of a properly equipped playground, while twenty years later Evelyn Sharp stated in her account of London childhoods that of all the thrills supplied in a modern playground, 'the greatest of them all is undoubtedly the sand-pit'.[124] In 1915, fourteen of the thirty-one play spaces managed by the LCC included a sandpit, while by the 1930s there was sand in forty-three of their forty-nine playgrounds.[125] Sandpits even featured in debates in the House of Commons, with the First Commissioner of Works, George Lansbury, stating that 'anyone who has seen children from the slum areas enjoying themselves in the sandpits will agree that the provision of these playgrounds has been a very good thing indeed'.[126]

At the same time, the place of sand in the playground was not straightforward. The popularity of sandpits among park users, and most likely a degree of class prejudice, meant that sand was often seen as problematic by park authorities. By 1909, parks staff were disinfecting the Kensington Gardens sandpit with permanganate of potash once a week, turning the sand over once a month and replacing it entirely once a year.[127] Commercial versions of permanganate of potash, such as Condy's Fluid, were promoted for their ability to prevent the spread of infectious diseases and purify crowded places, hinting at the concerns that park staff had about the potentially pathogenic nature of sand. The sandpit also challenged established expectations about park aesthetics and the forms of nature that were welcome there. In 1930, the St James's Park sandpit appeared 'very trodden down and untidy' and park managers felt that more frequent upkeep was needed.[128] The improved maintenance regime would make the park appear more respectable, while also apparently helping to deter fleas. It is not clear what type of fleas the authorities were concerned about (sand fleas, for example, live in the tropical and subtropical areas of America and Africa) nor whether fleas were actually a problem, but increased maintenance would deter this undesirable fauna none the less. Park superintendent Pettigrew went even further, suggesting that proper maintenance was only one part of ensuring a sandpit was safe. He argued that they also needed to be 'fenced in and only open to children when a play leader (preferably a young woman) is present to look after them' and that without these additional arrangements, the sandpit 'might easily become a menace to their health'.[129]

Designing the perfect play experience

While the hazards hidden in the playground underlined the contested meanings of healthy spaces, debates also continued about the form and function of the playground. The sandpit may have become an essential feature, but wider approaches to the design and layout of playgrounds were being challenged too. As we have seen, advocates of play leadership tended to emphasise adult involvement rather than provision of equipment as the key feature of a successful play space. In addition, some campaigners, town planners and noted designers were also beginning to question the predominance of manufactured play equipment in playgrounds, and the associated lack of beauty and exclusion of nature.

During the 1920s, nature and exercise had appeared together less frequently in campaigners' rhetoric, as organisations increasingly focused on promoting one or the other. As a notable advocate of children's physical exercise, the NPFA's rhetoric rarely promoted the benefits of closer interaction with the natural environment, beyond a general sense that fresh air was important to children's health.[130] In contrast, organisations such as the London Children's Garden Fund went beyond the benefits of fresh air by providing opportunities for children to interact closely with nature, in some ways taking on the mantle of helping poor children to experience curated forms of the natural world in the city. However, this separation of nature and play was short-lived and playground campaigners and landscape architects once more emphasised the need to include space for both exercise and interaction with nature in parks and play spaces.

As someone involved in advising on the details of play space creation, NPFA committee member Commander Coote embodied many of these ideas and demonstrated how they influenced the practice of providing for children's play. He was critical of earlier versions of the playground, both in its 'levelled and gravelled' and 'equipped' forms. He felt that 'there have been too many unattractive asphalt areas, congested with apparatus' and instead argued that the 'children's playground should be no less beautiful to look at than a well-kept bowling green'.[131] For Coote, equipment and the wider playground should be attractive as well as functional. That said, he was not calling for the removal of apparatus from the ideal playground, something that would have been incongruous, given that he had seemingly worked with play equipment manufacturers Spencer, Heath and George to design the 'Commander B. T. Coote Model Combination Climbing Frame'.[132] Instead, he called for 'natural beauty' to be incorporated into play spaces as well. In many ways this echoed the work of the MPGA in the 1880s and 1890s, when they attempted to provide gardens where children could experience physical activity and interaction with natural features, but with adaptations for the modern world. In his vision of the playground, the limitations of equipment and in particular the apparent monotony of swings and slides would be supplemented by nature. Playground equipment would be combined with grass, flowers, shrubs and trees, ponds, sandpits and fountains. At the same time, a play leader would 'make the playground a magnetic force to draw the children away from the dangers and excitements of the streets'.[133]

This vision of the playground received a further boost – and a specific moniker – with the publication of *PlayParks* by the Coronation Planting Committee in 1937. The committee had been established to promote horticultural and arboricultural celebrations for George VI's coronation and a guide to the design of play spaces might therefore seem incongruous. However, the committee was chaired by landscape architect Marjory Allen and included representatives from a wide range of organisations that also supported the play space cause, including the NPFA, Institute of Park Administration, Town Planning Institute and the Garden Cities and

Town Planning Association. Although it undoubtedly focused on commemorative planting schemes, committee members also made use of this high-profile period of celebration to promote a revised vision for children's play spaces.[134]

PlayParks was written by Thomas Adams, the chair of the technical subcommittee of the Coronation Planting Committee. He was also president of the Institute of Landscape Architects and an influential town planner who had worked on Letchworth Garden City, planned the rebuilding of Halifax in Nova Scotia, initiated a planning system for Canada and planned the reconstruction and expansion of New York City.[135] In the foreword to *PlayParks*, the industrialist and philanthropist Lord Wakefield emphasised young children's 'right to play in the fresh air in perfect safety' and Adams went on to show how this could be achieved in practice, at the same time emphasising the importance of nature and staking a claim for the role of skilled designers in the creation of children's play spaces. If 'the street of tethered children', where parents secured young family members to posts with a length of rope, represented the problem for Adams, then the verdant lawns, trees and shrubs of a village green represented something of an imagined ideal. In contrast to the street and existing playgrounds which were often 'hard, bleak, and uninteresting', a well-equipped and well-organised playpark would provide a sense of liberty, helping to develop proper habits of play and promoting an appreciation of natural beauty.[136] In contrast to earlier conceptions of the playground, Adams emphasised that play spaces needed to foster gentle, imaginative and quiet play, as he felt there had been a tendency in the past to over-emphasise the benefits of energetic physical activity.

For Adams, playparks were as important as schools in the education of children, but they also offered wider benefits to society. As well as ensuring the proper physical and mental development of children, playparks would help to reduce crime, lessen noise nuisance, increase property values, contribute to the development of a civilised community, provide indirect economic value through a physically fit and happy workforce, and contribute towards the modernisation of rural areas. Adams felt that to realise this range of social benefits, a well-designed and properly supervised playpark would combine the best elements of the playground, park and garden. It needed natural features such as trees, rocks and pools to enable imaginary games like camping and hunting, attractive planting that would foster an appreciation of nature, as well as some playground equipment, which occupied the minimum space necessary. A playground most definitely did not need 'repulsive-looking fences' which Adams felt were a waste of money. In many ways, *PlayParks* represented a synthesis of advocates' thinking on dedicated play spaces, why society needed them, how they would benefit and nurture children, along with detailed considerations for their design.[137]

While the influence of the NPFA and other play space campaigners involved in the Coronation Planting Committee is evident, the publication of *PlayParks* can also be seen as a call for the greater involvement of skilled designers, notably landscape architects, in the design and layout of children's play spaces. In the 1920s and 1930s, individual landscape architects had been involved in the design and development of public parks. For example, the renowned landscape architect Thomas Mawson prepared designs for numerous public spaces around the world and his proposals often included large areas set aside for active recreation. Despite this, the wider landscape design profession seems to have rarely engaged with public park design, let alone spaces for children's play.[138] Mawson's son, Edward Prentice Mawson, seems to have been a solitary, and hardly prolific, exception. He had worked with Thomas Adams on *PlayParks*, preparing representative designs for large and small, urban and rural playparks, complete with planting, enhanced natural features and playground equipment. In 1935, Prentice Mawson also contributed an article to the trade journal *Parks, Golf Courses and Sports Grounds* where he emphasised the importance of parks and children's playgrounds for promoting public health, physical fitness and national prestige.[139]

Despite the efforts of the Mawson family, parks and playgrounds were of little interest to the wider landscape profession, perhaps reflecting an antipathy among designers towards landscapes for the masses and public spaces for children in particular. The Institute of Landscape Architects (ILA) was formed in 1929 and it was only after the publication of *PlayParks* in 1937 that children's play spaces were mentioned in its quarterly journal, *Landscape and Garden*. Even then, it was in a begrudging tone and sought to minimise their impact on the landscape. The first mention came in the summer of 1938 in the text of a lecture that a Captain J.D. O'Kelly gave to the ILA.[140] O'Kelly briefly mentioned children's play spaces alongside a commentary on recreational facilities more generally, noting that space for children's activities should be properly planned, well planted, attractive to the child, but also isolated within the park to limit noise nuisance to other park users.

A second mention came in spring 1939 when W.R. Pertwee contributed an article entitled 'Designing Children's Gymnasia'. While recognising the irrefutable arguments in favour of the need for play spaces, Pertwee felt that their appearance left much to be desired and suggested they were the least satisfactory feature of public parks. Pertwee's use of the older term 'gymnasia' perhaps reflected a last glance back to the nineteenth-century conception of the playground as space for energetic exercise, or it may have been a product of contemporary concerns with the physical fitness of the nation and a widespread interest in physical culture during the 1930s. Either way, Pertwee's detailed design advice was certainly not child-centric. It included suggestions to plant horse chestnuts to dampen the noise made by children, the use of berberis shrubs in place of fencing because their thorns were 'quite effective against children' and only a limited quantity of equipment to ensure children had enough space to queue up and await their turn.[141] If any interwar landscape architects were thinking about the provision of play spaces, they did not imagine that these spaces would take centre stage. Rather, an air of architectural elitism meant playgrounds should be hidden, protected from destructive children, and the visual and aural disturbance they created needed to be minimised.

While few landscape designers may have engaged with the issue of play space design, other professions were much more willing to discuss the purpose and form of playgrounds (and keen to host adverts from commercial playground equipment suppliers too). A well-organised and vocal parks profession had developed by the 1930s. The Institute for Park Administration was formed in 1926 and its monthly magazine, the *Journal of Park Administration, Horticulture and Recreation*, was involved in promoting play spaces from its very first edition in June 1936. The first page of this first edition was a full-page advert for the playground equipment of Charles Wicksteed & Co., including a photo of three spectacular-looking slides. The following month's edition included several articles which offered advice on playground layout and another which espoused the benefits of play leadership. After that, every edition included adverts for playground equipment manufacturers. From April 1937 until well after the Second World War, the entire front cover was devoted to a Wicksteed advert, while regular articles considered the role of the playground in relation to a range of topics including public health, slum clearance, town planning, wartime child evacuation from cities and juvenile delinquency.

These articles emphasised the role of the playground in remedying the failures of the past, tackling the problems of modern life, as well as offering hope for the future. At the same time, it seems likely that the commercial viability of *Park Administration* was at least in part dependent on advertising income. Editors had to balance the opinions expressed in the journal with the expectations of advertisers and as a result criticism of the playground was invariably limited to a call for more planting or a greater sensitivity to their location. By 1936, the editor felt that the problem of providing appropriate playgrounds 'which will really please children and at the same time prevent them from breaking their precious young necks' had been solved 'as far as humanly possible' by 'those two kindly wizards, Charles Wicksteed & Co. and B. Hirst & Sons'.[142] Despite the efforts of landscape architects to reintroduce nature into the playground, the parks trade press played a significant role in promoting and reinforcing a vision for the playground centred on manufactured playground equipment.

This chapter has explored how the National Playing Fields Association became an important advocate of dedicated places for children to play in the 1920s and 1930s. Through its lobbying, funding, publicity and guidance it helped to shape contemporary ideas about what a playground was for, how it should be designed and how adults should be involved in its use. At times it endorsed conservative social values, but it also highlighted the importance of commercial technologies and rational planning in shaping a better future for children and the nation. Its national coverage and sense of authority contributed to the standardisation of playground spaces across the country, while its day-to-day activities at a local level point to the ongoing significance of voluntary and municipal civic action in the interwar period. At the same time, as society increasingly claimed that streets belonged to motorists, the NPFA successfully promoted the playground as a key tool in protecting children from life-threatening aspects of the modern world and actively campaigned against alternative approaches that sought to regulate the city rather than children. As a result, the number of children's playgrounds increased significantly and a new orthodoxy, centred on manufactured swings in particular, became more firmly established in the minds of municipal administrators. By the 1930s, the London County Council managed nearly fifty equipped playgrounds and authorities in Manchester administered twenty-six, with many more provided by municipalities elsewhere.[143] The conviction that the playground was a space of health, education and safety obscured the threat to children from accidents, incidents and other maladies. The NPFA continued to operate during the Second World War, campaigning for play spaces for evacuees and lobbying for measures to protect children playing on the street during blackouts. The debates that it facilitated in the interwar years around play leadership, in conjunction with wartime damage to urban areas, would create an atmosphere that was conducive to the formation of postwar adventure playgrounds and a greater engagement by design professionals in the spaces where children were meant to play.

Notes

1. Edward Prentice Mawson, 'Public Parks and Playgrounds: A New Conception', *Parks, Golf Courses and Sports Grounds*, 1 (1935), 7–8.

2. W.W. Pettigrew, *Municipal Parks: Layout, Management and Administration* (London: Journal of Park Administration, 1937), RHS Lindley, 969.2 PET; W.W. Pettigrew, 'The Manchester and Salford Gardens Guild', *Radio Times*, 14 October 1927, p. 78, Radio Times Archive; W.W. Pettigrew, 'The Northern Garden', *Radio Times*, 24 April 1931, p. 243, Radio Times Archive.

3. Tom Hulme, 'Putting the City Back into Citizenship: Civics Education and Local Government in Britain, 1918–45', *Twentieth Century British History*, 26.1 (2015), 26–51; Charlotte Wildman, *Urban Redevelopment and Modernity in Liverpool and Manchester, 1918–1939* (London: Bloomsbury, 2016).

4. The London Playing Fields committee was renamed the London Playing Fields Society in 1899, 'London Playing Fields Society', *The Times*, 3 June 1899, p. 14; 'Play Fields for Youths: New Manchester Movement', *Manchester Courier*, 9 July 1907, p. 9; 'The Playing Fields Society', *The Manchester Guardian*, 19 October 1908, p. 10; E. Chandos Leigh, 'The London Playing Fields Society', *The Times*, 20 January 1911, p. 19; Basil Holmes, 'More Playing Fields', *The Times*, 12 March 1920, p. 12.

5. 'Playing Fields for the People', *The Manchester Guardian*, 4 April 1925, p. 7; 'Playing-Grounds for Young People', *The Times*, 4 April 1925, p. 17.

6. Carole O'Reilly, 'From "The People" to "The Citizen": The Emergence of the Edwardian Municipal Park in Manchester, 1902–1912', *Urban History*, 40 (2013), 136–55.

7. National Playing Fields Association, 'Draft Constitution', n.d., London Metropolitan Archive, LCC/CL/PK/01/042.

8. National Playing Fields Association, *First Annual Report* (National Playing Fields Association, 1927), p. 12, National Archives, CB 4/1.

9. National Playing Fields Association, 'Report of Proceedings at Inaugural Meeting of the National Playing Fields Association', 1925, National Archives, CB 1/1; National Playing Fields Association, *Second Annual Report* (National Playing Fields Association, 1928), p. 20, National Archives, CB 4/1.

10. Peter Borsay, *A History of Leisure: The British Experience since 1500* (Basingstoke: Palgrave Macmillan, 2006).

11. National Playing Fields Association, *First Annual Report*, pp. 3–8; 'The Playing Fields Association', *The Manchester Guardian*, 25 July 1925, p. 7.

12. Elizabeth Baigent, 'Chubb, Sir Lawrence Wensley (1873–1948), Environmental Campaigner', *Oxford Dictionary of National Biography* (Oxford: Oxford University Press, 2004).

13. National Playing Fields Association, *Third Annual Report* (National Playing Fields Association, 1929), p. 25, National Archives, CB 4/1.

14. National Playing Fields Association, *First Annual Report*, p. 35.

15. National Playing Fields Association, *Second Annual Report*, p. 16.

16. Tom Hulme, *After the Shock City: Urban Culture and the Making of Modern Citizenship* (Woodbridge: Boydell Press, 2019).

17. National Playing Fields Association, *First Annual Report*, pp. 15–16.

18. 'Speeches from the Meeting Held by the National Playing Fields Association', *Radio Times*, 3 July 1925, p. 57, Radio Times Archive.

19. Raphael Jackson, 'The Week's Good Cause: Appeal on Behalf of the National Playing Fields Association', *Radio Times*, 5 June 1936, p. 18, Radio Times Archive; M.W. Montgomery, 'The Week's Good Cause: Appeal on Behalf of the Glasgow and District Playing-Fields Association', *Radio Times*, 24 May 1929, p. 403, Radio Times Archive; T.H. Vile, 'The Week's Good Cause: Appeal on Behalf of the Monmouthshire Branch of the National Playing Fields Association', *Radio Times*, 8 July 1938, p. 25, Radio Times Archive; Tom Voyce, 'The Week's Good Cause: Appeal on Behalf of the Gloucestershire Playing Fields Association', *Radio Times*, 12 February 1937, p. 22, Radio Times Archive.

20. 'Need for Village Playing Fields', *Playing Fields Journal*, 2.7 (1934), 312.

21. Marjory Allen, 'The Coronation and the Village', *The Spectator*, 12 March 1937, p. 467.

22. P. Maud, 'Recreation in Public Parks and Open Spaces', *Playing Fields Journal*, 1.2 (1930), 7–13; B.T. Coote, 'Children's Playgrounds: Their Equipment and Use', *Journal of Park Administration, Horticulture and Recreation*, 1.2 (1936), 102–5.

23. London County Council, 'Recreational Facilities', 1935, London Metropolitan Archives, LCC/PK/GEN/02/003.

24. Claire Langhamer, *Women's Leisure in England 1920–60* (Manchester: Manchester University Press, 2000), p. 142.

25. Jean Williams, *A Game for Rough Girls? A History of Women's Football in Britain* (London: Routledge, 2003).

26. Richard Holt, *Sport and the British: A Modern History* (Oxford: Clarendon Press, 1992); Mike Huggins and Jack Williams, *Sport and the English 1918–1939* (London: Routledge, 2006).

27. J.A. Mangan, *'Manufactured' Masculinity: Making Imperial Manliness, Morality and Militarism* (London: Routledge, 2012).

28. Graham Dawson, *Soldier Heroes: British Adventure, Empire, and the Imagining of Masculinities* (London: Routledge, 1994).

29. Martin Francis, 'The Domestication of the Male? Recent Research on Nineteenth- and Twentieth-Century British Masculinity', *The Historical Journal*, 45 (2002), 637–52.

30. Melanie Tebbutt, *Being Boys: Youth, Leisure and Identity in the Inter-War Years* (Manchester: Manchester University Press, 2012).

31. Ian Beckett, Timothy Bowman and Mark Connelly, *The British Army and the First World War* (Cambridge: Cambridge University Press, 2017).

32. Prentice Mawson, 'Public Parks and Playgrounds', p. 7.

33. James Stark, *The Cult of Youth: Anti-Ageing in Modern Britain* (Cambridge: Cambridge University Press, 2020).

34. Ina Zweiniger-Bargielowska, 'Building a British Superman: Physical Culture in Interwar Britain', *Journal of Contemporary History*, 41 (2006), 595–610.

35. Anna Davin, *Growing up Poor* (London: Rivers Oram Press, 1996).

36. Bishop of Southwark, *House of Lords Debate, 14 February 1928, Vol. 70, Col. 92* (Hansard, 1928).

37. Keith Laybourn and David Taylor, 'Traffic Accidents and Road Safety: The Education of the Pedestrian and the Child, 1900–1970', in *The Battle for the Roads of Britain: Police, Motorists and the Law, c.1890s to 1970s*, ed. Keith Laybourn and David Taylor (London: Palgrave Macmillan, 2015), pp. 149–85.

38. 'Street Accidents to Children', *Playing Fields Journal*, 1.3 (1931), 19–20.

39. 'Street Accidents', *Playing Fields Journal*, 1.8 (1932), 12.

40. 'The Cry of the Children', *Playing Fields Journal*, 2.1 (1932), 4–5.

41. Board of Education and Ministry of Transport, *Report of the Inter-Departmental Committee (England & Wales) on Road Safety Among School Children* (London: HMSO, 1936).

42. G. Lowndes, 'The Cost of Traffic Accidents, the Case for Playing Fields from a New Angle', *Playing Fields Journal*, 4.1 (1936), 19–22.

43. Mabel Jane Reaney, 'The Urgent Need for Trained Play Leaders, a Paper given at the Conference on Play Leadership at the Institute of Education, London, 21 July', 1933, National Archives, CB 1/54.

44. Joe Moran, 'Crossing the Road in Britain, 1931–1976', *The Historical Journal*, 49 (2006), 477–96.

45. 'Street and Driving Accidents: The Coroner and Children's Playgrounds', *The Manchester Guardian*, 18 June 1907, p. 14.

46. 'The Heedless Child: What He Means to the Motorist', *The Manchester Guardian*, 9 October 1928, p. 8.

47. Metropolitan Police, 'Memorandum from "M" Division, Tower Bridge Station Regarding Juvenile Courts, 4 August', 1932, p. 2, National Archives, HO 45/15746.

48. Home Office, 'Metropolitan Juvenile Courts Statistics for the Year 1930', 1930, National Archives, HO 45/15746.

49. Nancy Astor, *House of Commons Debate, 28 April 1926, Vol.194, Col.2155* (Hansard, 1926).

50. 'Mothers Want Safer Streets', *Daily Mail*, 12 October 1937, p. 11.

51. 'Teaching Children Road Safety', *The Times*, 28 July 1938, p. 11; *Model Traffic Area No. 1* (Pathetones, 1938), British Pathé Archive, PT440 www.britishpathe.com/video/model-traffic-area-no-1 [accessed 6 July 2023]; H. Thornton Rutter, 'The Chronicle of the Car', *Illustrated London News*, 6 August 1938, p. 256.

52. 'Tottenham's Contribution to Road Safety', *Journal of Park Administration, Horticulture and Recreation*, 3.2 (1938), 95–6.

53. G.E. Paris, 'A Children's Playground and Model Traffic Area', *Playing Fields Journal*, 5.2 (1939), 89–96.

54. 'Training While They Play: Tottenham's Model Traffic Area', *Journal of Park Administration, Horticulture and Recreation*, 12.3 (1947), 85–8.

55. 'Dunfermline Trains Bairns in Road Sense', *Dundee Courier*, 28 June 1950, p. 2; 'Road Safety for Children', *The Manchester Guardian*, 1 May 1956, p. 14; 'Royal 12-Day Tour Begins', *The Times*, 28 June 1958, p. 4.

56. 'Traffic and Children: Expecting Too Much Wisdom', *The Manchester Guardian*, 2 October 1936, p. 8.

57. Krista Cowman, 'Play Streets: Women, Children and the Problem of Urban Traffic, 1930–1970', *Social History*, 42 (2017), 233–56.

58. F.R. Bush to Chief Commissioner of Police, letter, 13 February 1930, National Archives, MEPO 2/7803.

59. 'Children Revel in Street Playground', *New York Times*, 26 July 1914, p. 11; Will R. Reeves, 'Report of Committee on Street Play', *The Journal of Educational Sociology*, 4.10 (1931), 607–18.

60. Metropolitan Boroughs' Standing Joint Committee, 'Proposed Closing of Streets to Traffic for Use of Children', 1934, National Archives, MEPO 2/7803.

61. Board of Education and Ministry of Transport, *Road Safety Among School Children Report*.

62. Alfred Barnes, *House of Commons Debate, 28 March 1950, Vol.473, Col.33* (Hansard, 1950).

63. Lawrence Chubb to Ernest Holderness, 'Street Closures for Children's Play', 17 October 1932, National Archives, MEPO 2/7803.

64. 'The Use of Closed Streets as Playgrounds', *Playing Fields Journal*, 3.2 (1935), 35–6.

65. National Playing Fields Association, *Second Annual Report*, p. 12.

66. Lord Elgin, 'Letter to Sir Arthur Crosfield, Chairman of the NPFA', *NPFA Second Annual Report*, 1927, 12–13.

67. The NPFA allocated grants around the world on behalf of the King George's Field Foundation from 1935 to 1965, and administered the National Fitness Fund, created by the 1937 Physical Training and Recreation Act. National Playing Fields Association, 'King George's Fields County Register 1931–1965', 1965, National Archives, CB 2/24; National Playing Fields Association, 'Register of Grants Given under the Physical Training and Recreation Act 1937', 1942, National Archives, CB 1/83.

68. Prentice Mawson, 'Public Parks and Playgrounds', p. 7.

69. Noel Curtis-Bennett, 'Playing Fields in the Post-War Period', *Journal of Park Administration, Horticulture and Recreation*, 10.3 (1945), 47–63.

70. National Playing Fields Association, *First Annual Report*, p. 12.

71. 'Street Accidents to Children', p. 20; 'Children's Corner', *Playing Fields Journal*, 1.8 (1931), 29.

72. Maud, 'Recreation in Public Parks and Open Spaces', p. 13.

73. W.W. Pettigrew, *Handbook of Manchester Parks and Recreation Grounds* (Manchester: Manchester City Council, 1929), p. 14.

74. London County Council, 'Recreational Facilities'.

75. Adrian Wooldridge, *Measuring the Mind: Education and Psychology in England, c.1860 < c.1990* (Cambridge: Cambridge University Press, 1994), p. 113.

76. Carolyn Steedman, *Childhood, Culture and Class in Britain: Margaret McMillan, 1860–1931* (New Brunswick: Rutgers University Press, 1990), p. 101.

77. Margaret McMillan, *The Nursery School* (London: J. M. Dent & Sons, 1921), p. 47.

78. Philip Graham, *Susan Isaacs: A Life Freeing the Minds of Children* (London: Routledge, 2009).

79. Richard Bailey, *A.S. Neill* (London: Bloomsbury, 2013).

80. A.S. Neill, *Summerhill: A Radical Approach to Child-Rearing* (Harmondsworth: Penguin, 1968), p. 68.

81. Joseph H. Wicksteed, *The Challenge of Childhood: An Essay on Nature and Education* (London: Chapman & Hall, 1936), p. 116.

82. Burt is a controversial figure today. For a summary, see Pauline Mazumdar, 'Burt, Sir Cyril (1883–1971)', *Oxford Dictionary of National Biography* (Oxford: Oxford University Press, 2004).

83. Wooldridge, *Measuring the Mind*, p. 220; Jenny Willan, 'Revisiting Susan Isaacs – a Modern Educator for the Twenty-First Century', *International Journal of Early Years Education*, 17 (2009), 151–65.

84. Mabel Jane Reaney, 'A Director of Play', *The Manchester Guardian*, 11 March 1919, p. 4.

85. Mabel Jane Reaney, *The Place of Play in Education* (London: Methuen & Co., 1927).

86. National Playing Fields Association, 'Minutes of the Sub-Committee on Play Leadership 14 December', 1928, National Archives, CB 1/54.

87. W. John Morgan, 'The Miners' Welfare Fund in Britain 1920–1952', *Social Policy and Administration*, 24 (1990), 199–211; 'Miners' Welfare: A Model Playground in Wales', *The Times*, 2 June 1930, p. 11; *Lord Chelmsford Opens First Children's Playground to Be Erected under Miners' Welfare Fund in Colliery Districts*, 1926, British Pathé Archive, 634.18 https://www.britishpathe.com/video/lord-chelmsford-1 [accessed 6 July 2023].

88. National Playing Fields Association, 'Play Leadership in the United States', 1933, National Archives, CB 1/54.

89. National Playing Fields Association, 'Report of Conference on Play Leadership', 1933, pp. 7–14, National Archives, CB 1/54.

90. Reaney, 'The Urgent Need for Trained Play Leaders', pp. 5–6.

91. 'The Need for Play', *Hospital*, 66 (1919), 52.

92. National Playing Fields Association, *Sixth Annual Report* (National Playing Fields Association, 1932), pp. 11–19, National Archives, CB 4/1.

93. National Playing Fields Association, 'Minutes of Play Leadership Committee 23 April', 1936, National Archives, CB 1/54.

94. National Playing Fields Association, *Sixth Annual Report*, p. 19.

95. John Shepherd, 'Lansbury, George (1859–1940), Leader of the Labour Party', *Oxford Dictionary of National Biography* (Oxford: Oxford University Press, 2004).

96. 'Brighter London Parks: Mr. Lansbury Continues His Tour', *The Manchester Guardian*, 9 October 1929, p. 17.

97. 'Hyde Park Playing Field: £5,000 Offer to Mr. Lansbury', *The Manchester Guardian*, 1 November 1929, p. 12; Peter Donaldson, *Sport, War and the British* (London: Routledge, 2020).

98. 'Hyde Park or Coney Island?', *The Times*, 7 February 1930, p. 15.

99. 'Questions in the Commons: The Royal Parks: Mr. Lansbury's Plans Challenged', *The Manchester Guardian*, 18 February 1930, p. 6; John Pybus, *House of Commons Debate, Royal Parks and Pleasure Gardens*, 24 February 1930 *Vol.235 Col.1909* (Hansard, 1930); 'Mr. Lansbury and the Royal Parks: Criticism and Defence in Commons', *The Manchester Guardian*, 25 February 1930, p. 5.

100. 'Mr Lansbury and the Parks: The New Technique: Correspondents' Protests', *The Times*, 8 February 1930, p. 13.

101. 'Mr. Lansbury and the Parks: Hyde Park or Coney Island? Reply to Critics', *The Observer*, 9 February 1930, p. 17; 'Mr Lansbury and the London Parks', *The Manchester Guardian*, 11 February 1930, p. 6.

102. 'Mr Lansbury and the Parks: Peace or Playground: The Serpentine Pavilion', *The Times*, 12 February 1930, p. 15.

103. 'Fall of a School Bell: Child Killed in the Playground', *The Manchester Guardian*, 29 November 1906, p. 7; 'Explosion in a City Schoolyard: Eight Boys Injured, Accident during Repair of Cable', *The Manchester Guardian*, 29 May 1924, p. 9; 'Runaway Horse Kills Kid', *The Manchester Guardian*, 4 December 1924, p. 9.

104. 'The Swing Accident in an Edinburgh Playground', *Edinburgh Evening News*, 26 January 1905, p. 4; 'Corporations and Public Playgrounds: A Question of Liability', *The Manchester Guardian*, 10 January 1905, p. 12; 'Fatal Accident in Charlton Kings Playground', *Gloucester Citizen*, 19 December 1912, p. 5.

105. 'Success of "Safety First" Campaign', *The Times*, 29 March 1924, p. 15; C.G. Ingall, 'Industrial Accident Prevention', *The Journal of State Medicine*, 35.6 (1927), 360–65; Laybourn and Taylor, 'Traffic Accidents and Road Safety'.

106. Charles Wicksteed & Co., 'Playground Equipment, Tennis Posts, Fencing and Park Seats', 1926, p. 10, Wicksteed Park Archive, uncatalogued; Spencer, Heath and George Ltd, 'Playground Apparatus, 1932 Improved Models', *Playing Fields Journal*, 1.8 (1932), xiii.

107. 'Insuring against Third Party Risks: Playground Accidents', *Playing Fields Journal*, 1.3 (1931), 13.

108. 'A Recreation Ground Swing Accident', *Playing Fields Journal*, 2.6 (1934), 266.

109. 'Crown Court: £1,500 Damages for Boy, Playground Accident', *The Manchester Guardian*, 22 December 1936, p. 5; 'Child's Injury in Playground: Damages Doubled, Corporation Loses Appeal', *The Manchester Guardian*, 14 July 1937, p. 15.

110. Royal Parks, 'Memoranda in Response to an Article in the Times, Dated 18 November 1933, on a Recreation Ground Swing Accident in Walthamstow', 1933, National Archives, WORKS/16/846.

111. 'Court of Appeal', *The Times*, 15 March 1934, p. 4.

112. 'Playground Risks', *Playing Fields Journal*, 3.1 (1934), 3–4; Play Safety Forum, *Managing Risk in Play Provision: A Position Statement* (London: Children's Play Council, 2002); David Ball, Tim Gill and Bernard Spiegal, *Managing Risk in Play Provision: Implementation Guide* (London: National Children's Bureau, 2012).

113. Royal Parks, 'Park Keepers' Reports of Offences against Children', 1932, National Archives, WORKS/16/532.

114. Pettigrew, *Handbook of Manchester Parks*, p. 26.

115. London County Council Education Department to Bailiff of the Royal Parks, 'Child Molestation Cases', 27 January 1932, National Archives, WORKS/16/532.

116. London County Council, 'Playgrounds – Dry', 1927, London Metropolitan Archives, LCC/CL/PK/01/038.

117. John K. Walton, *The British Seaside: Holidays and Resorts in the Twentieth Century* (Manchester: Manchester University Press, 2000).

118. F.A. Boddy, 'Playgrounds for Children', *Journal of Park Administration, Horticulture and Recreation*, 3.11 (1939), 345.

119. Ann Marie Murnaghan and Laura Shillington, 'Digging Outside the Sandbox: Ecological Politics of Sand and Urban Children', in *Children, Nature, Cities*, ed. Ann Marie Murnaghan and Laura Shillington (London: Routledge, 2016).

120. Friedrich Froebel, *Froebel Letters: Edited with Explanatory Notes and Additional Matter by A. H. Heinemann* (Boston: Lee & Shepard, 1893).

121. Joseph Lee, 'Play for Home', *The Playground*, 6.5 (1912), 146–58 (p. 152).

122. London County Council, 'Report of the Parks and Open Spaces Committee, 16 May', 1893, p. 5, London Metropolitan Archives, LCC/CL/PK/01/104.

123. John James Sexby, *The Municipal Parks, Gardens, and Open Spaces of London: Their History and Associations* (London: Elliot Stock, 1898), p. 556.

124. Annesley Kenealy, 'Playgrounds in the Parks: A Plea for the Children', *The Daily Mail*, 14 March 1907, p. 6; Evelyn Sharp, *The London Child* (London: John Lane, 1927), p. 95.

125. London County Council, 'Parks and Open Spaces: Regulations Relating to Games, Together with Particulars of the Facilities Afforded for General Recreation', 1915, LSE Library, 421 (129D); London County Council, 'Recreational Facilities'.

126. George Lansbury, *House of Commons Debate, 24 February 1930, Vol.235, Col.1894* (Hansard, 1930).

127. Royal Parks, 'Kensington Gardens Children's Playground', 1909, National Archives, WORKS/16/391.

128. Royal Parks, 'St. James's Park. Children's Playground and Paddling Facilities', 1930, National Archives, WORKS/16/1504.

129. Pettigrew, *Municipal Parks*, p. 16.

130. David Niven, 'The Parks and Open Spaces of London', in *London of the Future*, ed. Aston Webb (London: The London Society, 1921), pp. 235–50.

131. Coote, 'Children's Playgrounds', p. 103.

132. Spencer, Heath and George Ltd, 'Playground Apparatus'.

133. Coote, 'Children's Playgrounds', p. 105.

134. Coronation Planting Committee, *The Royal Record of Tree Planting, the Provision of Open Spaces, Recreation Grounds & Other Schemes Undertaken in the British Empire and Elsewhere, Especially in the United States of America, in Honour of the Coronation of His Majesty King George VI* (Cambridge: Cambridge University Press, 1939).

135. Michael Simpson, 'Adams, Thomas (1871–1940), Town and Country Planner', *Oxford Dictionary of National Biography* (Oxford: Oxford University Press, 2004).

136. Thomas Adams, *PlayParks with Suggestions for Their Design, Equipment and Planting* (London: Coronation Planting Committee, 1937).

137. Adams, *PlayParks*, pp. 15–26.

138. Ian C. Laurie, 'Public Parks and Spaces', in *Fifty Years of Landscape Design 1934–84*, ed. Sheila Harvey and Stephen Rettig (London: The Landscape Press, 1985), pp. 63–78.

139. Prentice Mawson, 'Public Parks and Playgrounds'.

140. J.D. O'Kelly, 'Parks and Playgrounds', *Landscape and Garden*, 5 (1938), 94–5.

141. W.R. Pertwee, 'Designing Children's Gymnasia', *Landscape and Garden*, 6 (1939), 28–31 (p. 29).

142. 'Seen at the Public Health Exhibition', *Journal of Park Administration, Horticulture and Recreation*, 1 (1936), 247.

143. J. Richardson, 'A Century of Playing Fields Progress', *Playing Fields*, 6.4 (1946), 155–60.

Chapter 4

Orthodoxy and adventure: 'playgrounds are often as bleak as barrack squares and just as boring'

By the mid-twentieth century, playgrounds had been created in considerable numbers, while advocates and manufacturers had established an increasingly standardised playground ideal. Children's playgrounds were imagined as exciting and healthy spaces of leisure, as well as a refuge from the life-threatening dangers of the street. Whether in green spaces or on housing estates, the parks profession was largely responsible for providing and laying out play spaces. Used to purchasing specialist technical equipment like mowers or greenhouses, they quickly adopted a similar approach with the playground, procuring manufactured equipment from commercial suppliers. As a result, the equipped playground with its swings, slides and roundabouts became the orthodox image of the place where children should play. As we shall see, children's playgrounds, like public spaces in general, were affected by the Second World War, but the fundamental assumption that playgrounds were necessary remained powerful. The creation of the postwar welfare state set the context for developments in mid-century playground thinking and helped to embed the playground more firmly into visions of both urban childhood and the urban environment. As town planners sought to create more rational and hopeful urban landscapes, dedicated spaces for play proved to be an important component of renewed neighbourhoods and new towns. In exploring these issues, this chapter addresses the general omission of the playground from otherwise comprehensive accounts of child welfare and the architecture of the welfare state.[1] But while the principle of the playground became more firmly established after the war, the orthodox form was seen as increasingly problematic. For its critics, the asphalt surface and metal equipment meant there was little room for nature, while the provision of apparatus that solely facilitated physical activity was seen as inadequate in meeting the holistic developmental needs of children.

In some ways, postwar campaigners displayed considerable continuity with the rhetoric of earlier playground advocates. They emphasised the detrimental social consequences of street play, particularly the perceived relationship between the street and delinquency, and the educational and health benefits that interaction with appropriate forms of nature offered for urban children. In other ways, their campaign rhetoric differed significantly and amplified the conceptions of childhood and playground critiques that had first appeared in the 1930s. The playground was still imagined as a space of childhood health, but rather than simply focus on the promotion of physical exercise in the open air, campaigners expected the playground to support children's cognitive and emotional development and provide a wider range of 'natural' experiences. Greenery and planting remained one aspect of the natural world that they sought to recreate, but opportunities to interact with mud and sand, water and wood also became increasingly significant. The fun and excitement that Charles Wicksteed felt his playground equipment represented was replaced among postwar campaigners by an emphasis on the need for the playground to facilitate children's freedom, creativity and self-expression.

This chapter shows how the National Playing Fields Association (NPFA) continued to play a significant role in coordinating playground advocacy and sharing knowledge. But at the same time, it highlights the increasing influence of a wider range of individuals and professions on play space design and provision. In particular, the chapter shows how town planners routinely allocated specific sites for play in their visions for a modern urban environment and the playground became an essential and everyday feature of both imagined and realised new urban communities. At the same time, child welfare campaigners and sociologists increasingly emphasised the need for a greater sensitivity to the diverse play interests of children and their perceived developmental needs. Campaigners, architects and artists imagined new playground forms which promoted children's cognitive, emotional, physical and social development. Marjory Allen (Lady Allen of Hurtwood, 1897–1976) in particular became a figurehead for alternative visions of the playground. In considering these wider influences on the playground form and function, this chapter expands the spatial and temporal scope of existing historiography on the postwar adventure playground movement.

Orthodoxy consolidated: postwar planners and the playground

During the Second World War, the themes and practices that had dominated interwar playground discourse continued to be important. The NPFA sustained a close working relationship with the government, loaning money to the war effort and promoting the playground as a way to address child fatalities on the roads.[2] On a practical level, the Association continued to acquire or partially fund a small number of new play spaces, including in Ferryhill in County Durham and King's Somborne in Hampshire.[3] At the same time, the fear of substantial casualties as a result of wartime bombing prompted far-reaching responses and had implications for the playground. Plans had been developed in the 1930s, particularly by London County Council (LCC) in London, to remove children from large cities in the event of Britain's involvement in a looming conflict.[4] During the war, over a million children moved intermittently back and forth between urban centres and rural evacuation areas, apparently leaving city swings idle and play streets empty.[5]

But, at the same time, children who remained in the city often made use of damaged buildings and bomb sites as informal places for play. Often remembered fondly by adults who spent their childhood playing in the war-torn city, such activities also inadvertently created a powerful contrast of childhood innocence and wartime devastation, one that would have a lasting impact.[6] A number of scholars have charted the simultaneously unsettling and inspiring terrain of the bombsite and its influence on contemporary writers, filmmakers, artists and architects, a significance that is explored later in this chapter. But, more immediately, anxiety about play on bomb-damaged sites further fuelled calls to remove children from the city and provide better places to play. Commentators complained of children's use of bomb-damaged buildings and looting of empty homes, while Ministry of Health propaganda posters urged children away from bombsites and encouraged their families to leave the city altogether.[7] Save the Children Fund created air raid shelter play centres in large cities, while playground advocates promoted the need for appropriate play facilities in evacuation areas, where relationships between evacuees and host communities were often strained.[8] Some disgruntled rural residents drew on earlier notions of degeneration to complain of the 'stunted, misshapen creatures' being sent from poor urban neighbourhoods into clean homes in the country, the city children seemingly beyond redemption even when relocated to more bucolic surroundings.[9] Others were slightly more sympathetic and felt that providing playgrounds in the green spaces of reception areas would help to tackle the inappropriate behaviour of evacuee children, as well as contributing to the future health and fitness of the nation.[10]

In February 1940, a contributor to the *Journal of Park Administration* felt that the 'beauties of Nature leave [evacuee children] stone cold', resulting in mischief and damage in the green spaces of reception areas, and that playgrounds would provide a distraction that would help to prevent such hooliganism.[11] However, only a few months later, manufacturers such as Wicksteed & Co. had to adapt their production lines to war work and could no longer supply new equipment.[12] Parks were turned over to military installations and food growing, while children's playgrounds were impacted by the war too. In 1940, the Salford parks committee decided to remove iron railings and playground swings for war purposes, while the playground in Ardwick Green Park in Manchester was summarily requisitioned by the military, to the consternation of local open space advocates.[13] In London, Paddington Recreation Ground was repurposed as a municipal piggery, with pens made from bombed timber and food waste used as feed. In two years, the recreation ground supplied over 300 tonnes of pig meat and generated an annual net profit of over £320,000 (£2,250).[14]

The recreation ground was re-opened in May 1948 by Field Marshal Montgomery as a children's playground with manufactured equipment, paddling pool and sandpit.[15] In recreating a playground in this form, Paddington borough council's parks department drew upon a vision of the ideal playground that had been largely settled in the minds and practices of park administrators since the 1930s. Writing in 1946, the director of the Institute of Park Administration demonstrated the ongoing influence of this ideal type, suggesting that 'a children's paradise' should include manufactured equipment, unclimbable fencing and asphalt surfacing.[16] The images that he used to support his article demonstrate a rather sombre and sanitised vision of a childhood utopia, devoid of the garden planting seen in nineteenth-century children's gymnasiums or the apparent excitement of the Wicksteed Park playground. This certainty among park managers about the ideal form of the playground continued for at least the next ten years. During the late 1940s and early 1950s contributors to parks trade journals rarely mentioned children's playgrounds, other than to note their existence in accounts of public parks in cities such as Cardiff and Portsmouth. Postwar shortages of raw materials and skilled labour, as well as high inflation and a thirty-three per cent purchase tax, made the supply of playground equipment more problematic, but it did not dent the stability of the equipped playground ideal type.[17] Nor did it seem to affect the provision of play spaces on the ground. The NPFA provided funding for 1,313 playground projects in the ten years after the war, compared to the 1,017 in the fifteen years before the war.[18] The 'orthodox' playground had become a well-established and familiar part of the mid-century park superintendent's responsibilities, alongside nursing plants and mowing lawns.

With childhood wellbeing as a central tenet of the postwar welfare state, the place of the playground in wider social policy perhaps seemed even more secure. But while the health and wellbeing of children and their families were an important feature, the playground was not an explicit component of national welfare policies. The purported benefits of open-air schools and the curative power of open-air treatments for tuberculosis were increasingly being questioned, as cold and windswept conditions hindered children's education and clinical approaches to the treatment of TB evolved.[19] And while the playground ideal had long been premised on broader notions of health and education, playgrounds did not feature in legislation that created a national health service, provided financial support for the family or sought to maintain full employment. The 1944 Education Act imposed a duty on Local Education Authorities to provide adequate facilities for children's recreation, but it did not make the creation of public play spaces compulsory.[20]

Instead, postwar playground provision took place within a wider collectivist and universalist atmosphere in which there was a broad political consensus about the role of government in delivering social democratic policies and practical interventions in the urban environment.[21] Within this context, the egalitarian potential of the park and the social possibilities of the playground meant that spaces for play were given a boost by the values and objectives of the wider welfare state. At the same time, children's playgrounds remained the discretionary responsibility of local authorities, and municipal parks departments continued to provide playgrounds much as they had done since the 1920s, with emphasis remaining on the provision of manufactured play equipment. The creation of new spaces for play was meanwhile taken up by the planning profession, representing a major intervention in the story of the playground. In exploring the position of the children's playground in the postwar planning landscape, this section offers an important new account of town planners' role in consistently championing play space provision when imagining and creating new urban environments. Planners' involvement in advocating for and designating space for play was not new in 1945 and it represented an expansion of earlier efforts to solve the problem of playing in the city.

Attempts to rationally plan the urban environment had their roots in nineteenth-century efforts to tackle the chaotic and unhealthy consequences of industrialisation and urbanisation. Interested in the distribution of green space within towns and cities, planners and play space advocates had long sought to develop a more rational approach to the location of parks generally and playgrounds in particular. Since the 1890s, when Reginald Brabazon had called for children's playgrounds every half a mile in London's working-class neighbourhoods, prescribing the right amount and frequency of play space had been a consistent concern. The eminent town planner Sir George Pepler (1882–1959) similarly stressed the need for town planning to include the rational provision of dedicated spaces for children. In 1923 he proposed a standard requirement of one third of an acre of play space per thousand residents (he would later become an influential member of the NPFA, serving on its Council and several committees).[22] Later attempts to develop a standard for playground provision would emphasise an amount of play space per child, for example twenty-five square feet per young child, eighty square feet per older child.[23] Despite the attempts to establish standards for provision, a nationwide investigation overseen by Pepler in 1951 found that many local authorities provided less than 0.05 acres of dedicated play space per thousand residents, well short of the 0.3 acres Pepler had called for in 1923. At the same time, seventy-four per cent had no plans to increase play space provision. The investigation did not establish why so many had no intention to provide more playgrounds, but postwar austerity was perhaps a significant factor.[24]

The provision of public spaces, including playgrounds, had also been associated with attempts to address poor-quality housing. Nineteenth-century campaigners, such as Octavia Hill, had included space for play in experimental housing schemes, while one of the earliest council housing projects in London, the Boundary Estate (1900) in Bethnal Green, included a raised ornamental garden at the centre of its radiating streets. By the 1920s, several government committees had endorsed the need for play spaces close to new council houses.[25] Over time the state gradually assumed a greater role in the provision of housing, first in the 1930s and then again from the 1950s on a far larger scale following wartime damage.[26] In the 1930s there was considerable and highly politicised debate about what form such council housing should take.[27] Garden city advocates promoted low-rise, low-density housing with public and private gardens on the edge or beyond the city boundary, as we saw with Charles

Wicksteed and his garden suburb. Others, often inspired by European modernism, promoted multi-storey blocks of flats as a direct in-situ replacement for slum clearance areas within towns and cities. Most authorities and the public came down on the side of low-rise houses with gardens, but in practice that did not solve the problem of providing space for play.

Between the wars, over four million new suburban homes were built by local authorities as council housing and by private developers for sale.[28] But the associated increase in the provision of private gardens did not eliminate the need for public play spaces. For many middle-class families, the popular gardening press promoted order, taste and decorum in the back garden with few concessions to children's play, other than the suitably cautious use of the lawn.[29] For many working-class families moving into new suburban council estates, the garden often provided a practical space where food could be grown, rather than somewhere for children to play.[30] As a result, even though the domestic garden was sometimes seen as the ideal place for play, in practice there was still a need to dedicate public spaces for children's recreation.

Although the suburban house and garden may have been the preferred solution for many, multi-storey blocks of flats were also built in the centre of some cities. With little private open space, these developments often included communal play provision. At Kensal House (1937) in west London, the housing specialist Elizabeth Denby and architect Maxwell Fry designed a modern block of flats for the Gas Light and Coke Company and included a children's playground as a central feature of the wider 'urban village' amenities.[31] At White City (1939) the LCC housed 11,000 residents in five-storey blocks and the plans included two 'fitted playgrounds' but this time squeezed in on the periphery of the estate.[32] Elsewhere in London, playgrounds also featured in slum clearance schemes in Poplar and Deptford.[33]

The layout of play spaces on housing estates repeated that seen in public parks at the time, closely fitting the orthodox playground ideal. In 1939, Manchester's director of parks felt that the ideal play space for clearance schemes should comprise modern playground apparatus and a border of trees, shrubs and flowers.[34] Contemporary photos suggest that the former appeared more regularly than the latter. In Leeds, the strikingly modern Quarry Hill estate (1938) included a playground in the central courtyards created by the blocks of flats. Surrounded by fencing, the slide and other items of play equipment represented a good example of the interwar orthodox playground, while the substitution of grass lawns with asphalt further enhanced the modernist aesthetic.[35] In Liverpool, the Caryl Gardens tenement scheme (1937), built under the auspices of the city's director of housing Lancelot Keay, provided another good example. Keay was a distinguished municipal architect and later president of the Royal Institute of British Architects (RIBA) and a member of the Ministry of Health's housing advisory committee.[36] Strongly influenced by the ideas and aesthetic of European architectural modernism, he was a firm advocate of multi-storey housing and the children's playground formed a central feature of several of his schemes. At Caryl Gardens, the central play space matched the orthodox ideal, with plenty of swings installed on an asphalt surface, surrounded by fencing (Figure 4.1). Elsewhere in Liverpool, the St Andrew's Gardens scheme also included a courtyard playground, with manufactured equipment and seating made from ship's timbers.[37] The division of responsibility seen in these examples, with architect-planners designating space for play and park managers directing its form, would be a consistent feature of subsequent town planning too.

From inclusion in a handful of interwar housing schemes, the playground became a regular feature of more comprehensive planning that took place during the war. In 1941, the *Picture Post* ran a special edition entitled 'A Plan for Britain', promoting alternative ideas for postwar social welfare, housing, education and health. The children's playground could have featured in any of these spheres of public life, but it was in an

Figure 4.1: Children's playground, Caryl Garden Flats, Liverpool by J.E. Marsh, 1940, RIBA Collections, RIBA14445.

article by architect Maxwell Fry that the potential of the playground was stressed. Fry asserted that postwar Britain 'must be planned' and that one of the many urban evils that proper planning could tackle was the lack of children's playgrounds.[38] Fry's emphasis on play space provision can perhaps be ascribed in part to his earlier experience of working with Elizabeth Denby at Kensal House, as well as his business partner, Thomas Adams, who had written *PlayParks* in 1937. Just two years later, play space provision was a specific requirement set out in several official and unofficial planning documents which considered postwar reconstruction. In London, the Royal Academy's advisory plan mentioned a general need for children's play spaces, while the official *County of London Plan* (1943) set standards for the ideal distribution of green space and included Pepler's prewar recommendation for a third of an acre of play space for children.[39] The prominent town planner Patrick Abercrombie, who had shared the stage with Charles Wicksteed at the Leeds town planning conference in 1918, prepared the *County of London Plan*, along with J.H. Forshaw, architect to the LCC and formerly of the Miners' Welfare Committee.

Abercrombie would also include play space in planning proposals for other cities too. *A Plan for Plymouth* (1943), prepared with city engineer James Paton Watson, 'intended to cover the whole of its existence from the comfort and convenience of the smallest house and children's playground to the magnificence of its civic centre'.[40] The plan called for a system of playgrounds every quarter of a mile, located in public parks or on the sites of demolished housing, and suggested that an additional eighty-seven play spaces were needed on top of the existing eighteen. Paton Watson was responsible for the implementation of the plan and a commentator would later conclude that he had successfully created children's playgrounds on a scale not equalled by any other city in the country.[41] The implementation of the London plan also increased the amount of space allocated for play. For example, at Spa Fields playground in Finsbury a combination of wartime bomb damage, compulsory purchase of poor-quality buildings and the appropriation of road space allowed the play area to be doubled in size and the setting of the earlier Finsbury Health Centre (1938) improved.[42] These planning schemes sought to impose a modern order on the seemingly chaotic and war-torn city. Narrow streets and overcrowded homes would be replaced with modern commerce and housing, while purpose-built playgrounds would provide cleaner, safer and more salubrious alternatives to the informal play spaces of the street and bombsite.

This sense that playgrounds had become an essential part of modern town planning and modern communities was best demonstrated by the place of the playground in the 1951 Festival of Britain. As both a tonic for the nation after the war and an expression of what it meant to be modern and British, the festival is perhaps best known for the iconic attractions on London's South Bank, including the Skylon Tower and Royal Festival Hall, the graphic design of the 'Festival Style' and the Pleasure Gardens at Battersea.[43] George Pepler was a member of the festival's Council for Architecture, Town Planning and Building Research, one of two advisory councils established to guide festival planning (the other dealt with science).[44] And while the NPFA had hoped that its work might 'form a conspicuous feature of the Festival', the 1951 exhibitions did more to show that the playground had become a standard and, in many ways, unremarkable part of visions of the modern urban environment.[45]

Children were well catered for, particularly in the Pleasure Gardens, which included the Nestlé-sponsored crèche, Peter Pan railway and Punch and Judy shows. However, it was at the Live Architecture Exhibition in east London that the children's playground would be most conspicuously on display. Identified as an area for comprehensive redevelopment in the *County of London Plan*, the Architecture Exhibition encompassed the newly built Lansbury Estate (named after George Lansbury, former local MP and royal parks commissioner who had promoted playgrounds in the 1930s). The exhibition was intended as a demonstration of the transformative potential of building science and town planning and, alongside the modern low-rise housing, several children's playgrounds were created on the estate.[46]

While Pepler's involvement in the planning of the festival may have ensured the presence of estate playgrounds, they were hardly central to the design showcase. In the exhibition visitor guide, the key for the site map listed over thirty other features including demonstration homes, shopping precincts, churches, schools and an old people's home before mentioning the children's playgrounds, just before the lavatories and main exit.[47] The playground was presented as an uncontroversial, everyday necessity in a vision of the city that was both idealistic and pragmatic.[48] Optimistically, the Lansbury playgrounds demonstrated that a rationally planned version of the city could provide a better place for children to play. Pragmatically, they assumed an orthodox form based on that found in many public parks. The playground had become one component of what urban historian Simon Gunn has described as a banal urban modernism, one that emphasised functionality rather than the iconic.[49] This functionalism saw planners attempt to organise the city into areas for industry, transport, living and at a micro-scale playing, while also making use of conventional assumptions about play space form.

This approach was replicated in the planning and layout of many post-war new towns, where the children's playground was an essential but unremarkable feature, much like roads, homes and shopping precincts. In proposals for Crawley in Sussex, each neighbourhood was planned to have a children's playground alongside other community facilities such as allotments and public gardens, day nurseries and maternity clinics.[50] The children's playgrounds proposed for Knutsford in Cheshire were to be located on a footpath system that was physically separated from roads and motor vehicles, while the plans for the creation of a new town at Rainhill in Merseyside included among its 'community equipment' a playground at the centre of each new housing area.[51] In 1951, Harlow New Town saw the completion of The Lawn; at ten storeys it was Britain's first residential tower block, and as the height of new housing increased elsewhere, so did concern among campaigners and the government about the provision of playgrounds around blocks of flats.

In the late 1940s, central government acknowledged the value of dedicated play spaces but hardly in emphatic terms. Its 1949 *Housing Manual* recognised that redevelopment provided an opportunity to improve play space provision, but simply suggested that 'reasonably accessible playgrounds' might be provided.[52] A few years later it sought evidence to support a standardised approach in new social housing developments. In 1952, the Ministry of Housing contacted the NPFA to request information on the issue of play provision for multi-storey housing, with a view to establishing a national play space standard specifically for high-rise residential buildings.[53] In response, George Pepler chaired a newly formed Children's Playground Technical Subcommittee, which quizzed the city architects of Birmingham, Leeds, Liverpool and the LCC and studied some of the existing playgrounds provided for flats in London. The subcommittee's research led the NPFA to publish *Playgrounds for Blocks of Flats* in 1953.[54] It reiterated earlier rhetoric about the dangers of the streets and presented children-without-a-playground as a threat to ornamental lawns, a menace to motorists and prone to juvenile delinquency. The report's researchers found that a considerable number of blocks had no playground provision and where space had been set aside it was inadequate – unimaginative, poorly designed and badly maintained. Of the ninety-six play spaces studied, seventy-three were surfaced with asphalt, eighty-nine were enclosed by fencing, fifty had equipment and only nine had any plants or trees. This was an explicit critique of the playgrounds that had been provided at Quarry Hill, Caryl Gardens and elsewhere.

These findings did not mean that the principle of the playground needed to be revisited, but rather that the playground form needed a new approach. In many ways *Playgrounds for Blocks of Flats* demonstrated a considerable degree of continuity in the criticisms of the orthodox, equipped playground first levelled in 1937 by Thomas Adams's *PlayParks*. Both encouraged the involvement of specialists with knowledge of children and garden design. Both called for more nature, in curated forms at least, in the shape of hills and valleys, trees and shrubs. The most significant difference that had occurred in the time between the two publications was that by the 1950s, Marjory Allen was leading an increasingly high-profile public campaign to improve play space provision, something explored in more detail later in this chapter.

Playgrounds for Blocks of Flats also laid bare the class, age and gender assumptions of the playground movement of the time. While blocks of flats were primarily being built as working-class social housing, the publication suggested that, along with other amenities, courts should be provided for the game 'fives', which was played almost exclusively at fee-paying public schools.[55] It also demonstrated particular expectations of the mothers who lived on estates, both in terms of assigning responsibility for the safety of children to them and the expectation that their work would be primarily based in the home. Play spaces for the youngest children, aged between two and five, were to be 'within sight of mothers in the flats' and enclosed with a self-closing gate, so that 'toddlers can be left on their own for short periods while their mothers get on with their work'.[56] Once children were older than five, their playgrounds no longer needed to be within sight of the flats, and by the time children were nine or older, the report suggested that playgrounds should be well away from dwellings. First published in 1953, *Playgrounds for Blocks of Flats* presciently pre-empted the rapid increase in the construction of tall blocks facilitated by the 1956 Housing Act. High-rise construction boomed in the late 1950s and 1960s, so that schemes of ten or more floors were increasingly common across Britain. However, the publication's recommendations were dismissed as too costly in central government's 1957 *Housing Handbook* and subsequently ignored in the design and development of many high-rise housing schemes.[57] As a result, the provision of play spaces for children living specifically in flats remained an ongoing issue for campaigners well into the 1970s.[58]

During the 1940s and early 1950s, both the principle and ideal form of the playground were consolidated in planning documents and subsequent redevelopment schemes. Planners emphasised the place of the playground in creating modern, functional and humane communities, while the orthodox, equipped playground was well established among parks professionals as the ideal form for both public parks and new housing estates. In addition, the wider, child-centred welfare consensus prioritised the health, education and general wellbeing of children. At the same time, it was these changing conceptions of childhood and new urban environments that would fuel increasingly high-profile criticism of the orthodox playground. Popular acceptance of the creative and emotional needs of children contrasted sharply with playgrounds that were still designed and laid out to promote physical exercise and were regularly devoid of any 'natural' features. This orthodoxy would be increasingly challenged during the 1950s and 1960s.

Marjory Allen and the challenge of adventure

From the mid-1950s, the purpose and form of the playground were being questioned with increasing urgency. On the one hand, campaigners emphasised the need to support greater creativity and self-expression among children, promoting less rigid and more adventurous play opportunities, where adults designated space for play but children were able to shape the detailed form. On the other hand, a number of professional designers were increasingly interested in creating play spaces for children, bringing adult creativity and imagination to bear on the playground. This section considers the centrality of Marjory Allen to these processes, her widely credited association with the junk playground movement and the seldom-acknowledged significance of her environmental consciousness. It goes on to explore the contribution made by professional designers to both the imagined and physical form of dedicated play spaces for children.

From the late 1940s through to the late 1960s, Marjory Allen was the most high-profile figure associated with attempts to reinvigorate the playground. As the next section shows, she is popularly associated with the introduction of 'junk playgrounds' to the UK in 1946, while scholarly accounts have rightly emphasised the effectiveness of her campaigning and her role in shaping postwar attitudes to childhood. Here, however, Allen is situated within the longer history and geography of playground thought and her unacknowledged environmental biography is considered.

Allen spent much of her early childhood on her family's rural smallholding. She subsequently attended the progressive Bedales school, worked as a gardener, studied horticulture and later married Clifford Allen, the Independent Labour Party politician, peace campaigner and from 1932 Lord Allen of Hurtwood.[59] Their daughter Polly attended a progressive nursery school, prompting Allen's interest in infant education and membership of the Nursery School Association.[60] She later became chair and then president of the Association, emulating Margaret McMillan who was the organisation's first president.[61] Allen felt that nursery education provided children with 'free space, fresh air, sunlight, companionship, and engrossing occupation', all formative features of her own rural childhood.[62] She was also profoundly influenced by the thinking of her close friend Elizabeth Denby. In 1938, Denby had concluded that when playgrounds were 'merely an expanse of tarmac or concrete, the damage to the children is almost criminal. All sensibility must be stifled in the ugly atmosphere of such barrack yards'.[63] This sentiment would be repeated by Allen in a later letter to *The Times* when she stated that 'municipal playgrounds are often as bleak as barrack squares and just as boring'.[64] Perhaps surprisingly, Allen's first high-profile intervention in public discourse was not related to the playground and instead focused on the plight of children in care.

In the early 1940s she was the figurehead for a campaign seeking better standards for children living in residential care homes. In her letter to *The Times* that brought this issue vividly into the public imagination, she described how 'children are being brought up under repressive conditions that are generations out of date' and argued that the needs of individual children were being ignored.[65] Allen's profile and contacts with politicians and the media meant that the campaign was highly effective and influential in shaping the 1948 Children's Act. She became a 'public placeholder' for critiques of the care system, a high-profile embodiment of ideas that had largely been previously developed and publicly expressed by others.[66] Allen's approach to campaigning and her unintentional position as public placeholder would be repeated in her subsequent playground advocacy work. Several scholars have argued that it was this public and persistent campaigning, political and social contacts, organisational skills and strong sense of purpose – rather than detailed technical knowledge – that was most significant in explaining her profile within the postwar playground movement.[67] In her memoirs, Allen recalled that her 'mania for keeping things moving' had been highly effective.[68] However, none of these accounts consider the importance of Allen's assumptions about nature and the way they interacted with her ideas about childhood, nor how this in turn shaped her approach to play provision.

Allen's interest in nature preceded her involvement in child welfare campaigns. After leaving school, Allen worked as a gardener and then studied horticulture at the University of Reading, at a time when only twenty per cent of university students were women and there was considerable opposition to greater equality at Reading in particular.[69] In time, Allen became a founding Fellow of the Institute of Landscape Architects (1929), along with the noted designers Thomas Mawson, Brenda Colvin and Richard Sudell.[70] Although she initially worked mainly on private commissions, she also had a wider interest in public green spaces and from 1937 to 1939 she chaired the Coronation Planting Committee which among its other activities published Thomas Adams's *PlayParks*. In addition to her professional influences, Allen would later recall her own childhood interactions with nature and contrast them with the lives of city children. Just like earlier campaigners, her commitment to recreating a version of her own bucolic childhood and her perception of the problems of the city are clear from her writing. In her autobiography, written in 1975, she recalled a romantic vision of her rural upbringing, close to nature and with freedom to be creative and imaginative:

> The wonderful and simple life of haymaking, milking cows, growing flowers and vegetables and learning the craft of making butter and cheese, and all the lovely sights and scents of the country, remain for me the most enduring memories of my life. When, later, I worked among children condemned to live in barbaric and sub-human city surroundings, my thoughts always returned to my early good fortune. The remembrance has made me more determined than ever to restore to these children some part of their lost childhood: gardens where they can keep their pets and enjoy their hobbies and perhaps watch their fathers working with real tools; secret places where they can create their own worlds; the shadow and mystery that lend enchantment to play ... Our active life in the Kentish countryside gave us these moments of wonder and awe.[71]

This emphasis on restoring nature to the city initially extended beyond childhood and Allen sought to provide 'moments of wonder' for adults too. Her early landscape design work focused on greening city buildings, including roof-top gardens for Selfridges department store and a block of flats for the St Pancras Housing Improvement Society, while as author of the *Manchester Guardian*'s Country Diary column she brought accounts of the country into the homes of urban readers. In addition, she promoted physical forms of nature close to the homes of working-class city dwellers. Flowerboxes incorporated into 'cheerful outside balconies' on new blocks of flats would provide space 'where the young baby can sleep ... the small child may rest in the air and the sun' and bring brightness and cheer to adult residents and the wider community.[72] In some ways, Allen's work echoed earlier efforts to bring aspects of nature into the city, where fresh air and interaction with curated forms of flora and fauna would provide an uplifting physical and moral influence on city dwellers.

Despite Allen's commitment to recreating an urban version of the nature she experienced in the countryside of southern England, her environmental consciousness has rarely featured in popular or scholarly accounts of her work. This is probably due to the timing of her first and most high-profile intervention into playground discourse (shortly after her successful campaigning for the Children's Act) and because the form of her intervention – a photo essay in the *Picture Post* magazine – displayed little evidence of the natural world. Instead, the essay took advantage of the opportunity offered by bomb-damaged sites to reimagine spaces for play and spoke more to changing conceptions of childhood than to ideas about the country in the city. And rather than emphasise the need for greener places to play, Allen focused on the features that she felt comprised a 'natural' childhood – self-expression, freedom, creativity, shadow and mystery – in contrast to 'unnatural' urban childhoods. Allen emphasised the qualities that she had experienced in her rural upbringing, including independence and imagination, rather than necessarily interaction with the sights and sounds of the country. While these qualities have seldom been emphasised in recent accounts of her campaigning work, they did have a considerable influence on her conception of the playground and her subsequent activism. Making sense of this requires an examination of her *Picture Post* essay and its position within the longer and wider history of playground thought.

Allen's most public contribution to playground discourse came in 1946. On a trip to Norway, Allen's flight stopped to refuel in Copenhagen, and she briefly visited a junk playground that had been created during the war on the Emdrup housing estate. Later that year, she published her much celebrated photo essay in the *Picture Post*, including striking photos and a vivid description of the Emdrup playground, titled 'Why Not Use Our Bombsites Like This?' The compelling images showed children building with scrap wood, digging in the mud, tending handmade structures and nursing a fire. The playground did not include manufactured equipment, but instead had a variety of loose materials, including bricks, timber, earth and water, and a skilled adult play leader who could support children in their play.[73] Allen described what she saw at Emdrup as 'something quite new and full of possibilities'.[74] For Allen, it seemed to represent a radical new form of play space provision, a profound break from the asphalt and equipment of the orthodox playgrounds found in British parks and housing estates. However, while the photos that accompanied Allen's article presented a strikingly different approach to play space provision, the idea of playing with junk and the need for adult involvement in children's play had been circulating for many years, both in Britain and beyond.

In Denmark, the Emdrup playground represented a long-standing intellectual and practical collaboration between landscape architects and pedagogues. In fact, rather than using an area of waste ground as Allen suggested in her essay, the site had been specifically designated as a playground in line with local building regulations and at least in part designed and laid out by a landscape architect. Allen (and, based on her assertion, many others since then) credited the Danish landscape architect Carl Theodore Sørensen (1893–1979) with the invention of the junk playground. Through his writing, teaching and practice, Sørensen had a profound impact on landscape design in Denmark, although his work remains virtually unknown elsewhere.[75] As a result, it is useful to explore in more depth the background to Sørensen's junk playground idea and the cultural context that produced this apparently revolutionary design.

As its largest city, Copenhagen was a focus for early playground advocacy in Denmark. The first children's playground was created in 1881, the city's Playground Association was formed in 1891, and early play space tended to be 'rectangular and surrounded by shady trees' with gymnastic apparatus and sandboxes introduced from London and Berlin in 1908.[76] Sand became an essential and regular feature of public play space and was particularly associated with the pedagogue Hans Dragehjelm. He founded the Froebel Society in Demark (1902) and among other work published a scientific study on Children's Play in Sand in 1909, emphasising its value in the physical, emotional and imaginative development of children.[77]

By the 1930s, the average Copenhagen playground was an enclosed asphalt or gravel space with a sandbox, water tap, a few swings and seating for mothers. Sørensen and Dragehjelm cooperated on a number of play space designs in this period and reacted against this apparently austere play environment. Their writing emphasised the need for more nature in the playground, particularly in the form of adapted features of the Danish cultural landscape such as fields and meadows, forests and beaches, and they designed spaces with a greater emphasis on the needs and interests of individual children.[78] In many ways this echoed the earlier writing of the US play space advocate G. Stanley Hall, who felt that the field, forest, hill and shoreline were important spaces for both play education, while also highlighting the international exchange of ideas among European and north American playground advocates at this time.[79] This emphasis on national cultural landscapes also chimes with Allen's idealisation of the aesthetic and cultural values associated with the English countryside and the potential benefits for urban children.

More broadly, the children's playground was a significant feature of Danish city planning discourse, culminating in the 1939 Copenhagen Building Act which made playgrounds mandatory for new housing schemes in the capital, a requirement extended to other parts of the country in 1961.[80] Sørensen had previously asserted that 'children's playgrounds are the city's most important form of public plantation'.[81] But rather than list the equipment needed to furnish such a space, he emphasised the importance of its location close to the homes of children and its function as a site for independent play and self-education. He went on to ask whether 'we could try to design a kind of junk playground in suitable and fairly large areas, where the children would be allowed to use old cars, cardboard boxes, branches and such'.[82] In 1943, Sørensen and Dragehjelm were given an opportunity to do just that, being commissioned to create such a space on the new workers housing estate at Emdrup. The design for Emdrup partially resembled earlier versions of the Danish playground, rectangular in shape and surrounded by dense hedging on top of an earth embankment, with fencing hidden in the planting. But rather than gymnastic equipment or a defined sandbox, the central space is shown with few permanent installations. Instead of fixed equipment the design shows log piles and fallen trees, a replica sailing boat, caves dug into the perimeter embankment and stylised figures digging and camping. The designers assumed that such a space might not need adult supervision, but once opened the Emdrup housing association that managed the playground employed a play worker to supervise its use. The collaboration between Dragehjelm and Sørensen meant that the Froebel-inspired outdoor classroom had been transported into the public realm.

Another notable feature of the sketch design is the tipi, stockade and games of Cowboys and Indians, a seemingly incongruous presence in the suburbs of the Danish capital. But from the late eighteenth century both representations of native Americans and indigenous people themselves had invoked a mix of both fear and fascination among British, European and North American fairground audiences.[83] In 1899, the pioneering cinematographers Mitchell and Kenyon of Blackburn dramatised playing at Indians in a short film.[84] By the early twentieth century, social discourse in the USA increasingly associated native Americans with wilderness, natural purity and authenticity, particularly for Ernest Seaton and the Boy Scout movement. Seaton's *Handbook of Woodcraft Scouting* (1910), written with Robert Baden-Powell, founder of the Scouts in Britain, popularised the long-standing associations between social constructs of native Americans and children, with both often characterised as more closely connected to the natural world than modern adults.[85] The shift within the Scouting movement from inward-looking militarism to an international liberalism, and specifically the staging of the Second World Scout Jamboree in Demark in 1924, perhaps helped to inspire the work of Sørensen and Dragehjelm at Emdrup.[86]

There were also much earlier examples of similar approaches to children's play provision elsewhere in the world and even Sørensen rejected the idea that he invented the junk playground, instead suggesting it was a loosely articulated concept. Over twenty years earlier, an experimental play space in the USA had made use of junk and waste materials in children's play. In 1918, the Bureau of Educational Experiments in New York sought to recreate in the city the play opportunities that were understood to be available to country children (the Bureau had been established by the educator Lucy Sprague Mitchell, after a visit to innovative schools in London in 1912).[87] Bringing rural play opportunities into the city would help to ensure both the 'muscle development of little people' and provide for their creative and dramatic play, progressive ideas for the time. The resulting outdoor laboratory created by the Bureau included bricks, lumber, tools and a packing-box village, in addition to a sandbox and rudimentary gymnastic equipment. Unlike Emdrup, this was a private, educational setting rather than a public, park-like environment.[88]

Similar suggestions for children's play existed in Britain too. In 1915, the sociologist and town planner Patrick Geddes had expressed the view that wigwam building, cave digging and stream damming were the natural activities of boys (girls were more naturally inclined to sit on the grass in Geddes's view).[89] A few years later, Raymond Unwin, the prominent architect and town planner, suggested that the design of play spaces could make more of existing landscape features and loose materials: 'any bit of unevenness in the ground, a hole or a mound, an old fallen tree, a few bricks, or such accessories are very helpful for little children's play'.[90] And the progressive educationalist Margaret McMillan suggested in 1921 that a wild corner of the garden, including stones, scrap metal and old pots, was an important feature of spaces for children, giving them freedom to play as they wished.[91] Even Allen had previously made use of junk. She coordinated a wartime project in which conscientious objectors made over three million toys and items of furniture from salvaged timber for children's nurseries.[92] The need for more flexible spaces that supported imaginative and creative play with loose materials had been circulating for thirty years before the Emdrup junk playground was created.

The appointment of an adult to support children's play was not a revolutionary suggestion in 1946 either. In 1919, Mabel Reaney had called for a director of play and since 1928 the NPFA had been coordinating the campaign work of play leadership advocates. Throughout the 1930s, the NPFA publicised the need for play leadership in *Playing Fields*, organised conferences and training courses and held practical demonstrations. In April 1946, eight months before Allen's *Picture Post* essay, Stockport had introduced adult games wardens to help organise children's play in its parks.[93] Not only was the idea of play leadership not new, but far from being revolutionary, its practical implementation often reinforced conservative social values. Historian Krista Cowman has shown how the postwar junk playground movement – and the appointment of heroic male playleaders in particular – did much to maintain traditional gender assumptions about both children's play and the role that adults had in supervising it.[94]

In emphasising the revolutionary nature of what she saw, it might seem that Allen either failed to acknowledge or was not aware of earlier alternative visions of the playground. The equipped and asphalted playground had been criticised in the 1930s, campaigners had been calling for the involvement of play leaders in children's play for over twenty years and ideas about playing with junk had been circulating for even longer. Two years before Allen's *Picture Post* essay, the Conservative MP Edward Keeling had suggested in the House of Commons that bombed sites could be repurposed and designated as playgrounds while public parks remained commandeered for military purposes.[95] Furthermore, Save the Children Fund had successfully operated a number of play centres during the war, including in and around the former Camel pub in Bethnal Green, where children were given considerable autonomy in organising their activities and sympathetic adults were largely in the background.[96]

Although Allen's 1946 essay may not have presented entirely new ideas, it did resonate with wider public debate about the impact of the war on childhood and the involvement of children in urban reconstruction. Images of children in a ruined landscape chimed with iconic representations of the poor urban child playing in the war-damaged city that had a compelling (and complex) place in the British postwar imagination. In some ways, the images represented society's hopes for the future and emphasised the child's role as an agent in the spatial and cultural reconstruction of postwar cities. Children could figuratively and literally help to rebuild the city. At the same time, play had assumed a therapeutic function. Based on the theories of psychologists such as Anna Freud and the practical, wartime experience of Marie Paneth at the Branch Street play centre, play could operate as an antidote to children's experiences of violence and destruction. In the junk playground, the city could help to rebuild a childhood affected by war.[97]

At the same time, such spaces could represent dystopian visions of destruction and chaos, or even the realisation of these visions for those living nearby. Marcus Lipton, MP for Brixton, received a 'constant stream of letters coming in from people who unfortunately live very near these small bombed sites, complaining of the filthy garbage, rotting mattresses, dead cats and all sorts of other things'.[98] In York, Alderman Buckton, chair of the city's housing committee, stated that 'tenants on our estates do not want these types of playgrounds'.[99] The borough council in Bethnal Green went further and reversed its decision to establish a junk playground, applying instead to the NPFA for funding to surface and lay out an orthodox play space.[100]

Despite the dystopian connotations, Allen's *Picture Post* essay and the approach to play space that it represented was influential. The essay brought together the benefits of playing with junk, the potential of play leadership and critiques of orthodox play spaces, linked them to the opportunities presented by bombed sites and presented them in a highly visible and accessible form. Before Allen's essay was published, these ideas had not challenged the dominance of the unsupervised, orthodox playground ideal. Afterwards a number of organisations were inspired to formalise children's use of bombed sites for play, most notably in London. The International Voluntary Service for Peace, University Settlement movement, Save the Children Fund and local community groups were all involved in early efforts to create junk playgrounds. The first opened in Morden in London in 1948 and the idea spread to other towns and cities, including Liverpool, Crawley, Bristol and Grimsby. Thanks to Allen's advocacy, in time the NPFA provided a degree of national coordination and the sharing of knowledge and experience through conferences, publications and committees.[101] Junk playgrounds gradually received more widespread attention, even if their number did not increase significantly. The thousands of orthodox playgrounds created in cities, towns and villages across Britain far outnumbered the seventeen junk playgrounds that were opened between 1948 and 1960. Many of the initial experimental spaces were only open for a few years, reclaimed by landowners as temporary leases expired, including the Camberwell (1948–51), Clydesdale Road (1952–5) and Lollard Street (1955–9) playgrounds.

The apparent chaos and destruction of the junk playground contrasted sharply with the ideal vision of the public park and goes some way in explaining why the junk playground did not become more widespread. Despite Allen's high-profile *Picture Post* article, the junk playground received little coverage in the parks trade press for instance. After a single image of a tree house in 1948, which was captioned as a junk playground, it would be another five years before experimental play spaces received greater publicity among green space professionals. In addition, Allen had transported the junk playground idea from a particular social and cultural context and dropped it into postwar Britain. In Denmark, playgrounds were a mandatory part of new urban landscapes, designers and pedagogues had long worked together, and child-centred notions of play, particularly in the form of the sandpit as a tool for children's self-expression, had long been seen in public play spaces. In contrast, there was no legislative compulsion to provide playgrounds in Britain, the orthodox image of the playground was one dominated by manufactured equipment, and there tended to be little communication, let alone active collaboration, between local authority parks and education committees or professional practitioners.

The British and Danish spaces were noticeably different in their implementation too. The Emdrup site had been specifically designated for play, deliberately designed with its earth embankment, planting and a purpose-built building with toilets and other facilities. In Britain, early junk playgrounds opened on bomb-damaged sites that were temporarily available, they had little infrastructure beyond a boundary fence, rarely had planting or other natural features and usually managed with scavenged huts or sheds. Even the term 'junk' was problematic – Allen and George Pepler apparently decided over a lunch that 'adventure' was a better term to use, and it first appeared in her pamphlet *Adventure Playgrounds* published by the NPFA in 1953. In addition to short leases, funding was also a practical problem for British junk playgrounds. Lollard Street was managed by a diverse supervisory committee, including Allen, the LCC, NPFA and Lambeth borough council. By insisting on 'the utmost economy in capital expenditure' the committee made it very difficult to replicate the Emdrup playground with its purpose-built building, sturdy boundary and planting.[102]

In addition, the absence of an agreed-upon definition for junk or adventure playgrounds meant that coverage could include a wide range of play spaces. For Allen, this was not entirely positive, as largely conventional play spaces were inappropriately labelled adventure playgrounds. A 1960 account of two new adventure playgrounds in Liverpool, one at Whitley Gardens and another at Kirkdale Recreation Ground, seemed to justify Allen's concerns. The features listed included conventional play equipment, putting green, ornamental planting and a play lawn for babies.[103] This was very different from both the experimental spaces created on bombsites in the years after the war and Allen's hopes for more naturalistic play opportunities.

Given Allen's earlier landscape work, her focus on bombsites and junk at the expense of nature might seem out of character. However, it seems likely that Allen temporarily put nature to one side to take advantage of the opportunity presented by bomb-damaged sites and evolving notions of childhood. But in doing so, Allen did not abandon nature altogether and instead she expressed reservations about the lack of natural features on many junk playgrounds. In a later, veiled criticism, she was 'delighted to see trees and a stream' at the Southmead adventure playground in Bristol, as natural features were missing from similar spaces elsewhere.[104] In acknowledging such a presence, Allen seems to have overlooked uncultivated bombsite plants and animals. From accounts in the 1950s, as well as more recent scholarship, we get a sense that bombsites were often rapidly colonised once the dust had settled. In a 1953 'note on new ruins' the writer Rose Macaulay poetically described how bombed buildings soon had trees sprouting through empty window sockets, while rose-bay and fennel blossomed in broken walls.[105] More recently, the archaeologist Gabriel Moshenska has shown how the untamed ecology of postwar bombsites was a consistent feature of first-hand accounts of wartime childhood. Within a few weeks of an air raid there was invariably a healthy crop of weeds, including the rose-bay willowherb, while ponds provided habitat for dragonflies and other aquatic insects. The director of Kew Gardens even gave a public lecture on wildflowers on bombed sites, identifying over 150 different species.[106] Nature was present but not in the form that Allen valued or appreciated.

Much of Allen's later writing on children's playgrounds would emphasise the disconnect between children and nature evident in the orthodox playground. The assault on the natural environment triggered by playground creation meant that 'streams are hidden in the sewers, the hills and mounds are levelled out, the good earth is buried under concrete and the trees are certainly not for climbing'.[107] In creating orthodox playgrounds, the ruthless bulldozer, unrelenting asphalt machine and the expensive ironmonger destroyed all the natural landscape features, creating a sense of imprisonment and doom. Without opportunities to interact with the fundamental elements of nature, including water, earth and fire, Allen felt it was unsurprising that children would express their primitive instincts in ways that were problematic for individuals and wider society. Allen would go on to position more naturalistic play spaces as the best solution to this problem in a number of subsequent publications.

Following a trip to Sweden in 1954, Allen published an article in the trade press and a booklet entitled *Play Parks* as a way to promote alternative visions of the playground.[108] She praised various aspects of Swedish play space design, but the presence of nature is the most notable feature of the booklet. Alongside play leaders and a variety of moveable materials, she found hedges of flowering shrubs, undulating grass meadow areas, roughly constructed wooden animals, birchwood building blocks and generous sandpits where the sun, rain and air helped to keep the sand clean and wholesome. Echoing Adams's *PlayParks* from 1937, she concluded that 'every playground should have some of the characteristics of a park or garden. Planting is not a mere decoration, it is a part of the necessary equipment of a modern playground'.[109] Some features of the Emdrup playground were present in Allen's *Play Parks*, including a supportive adult and facilities to support children's self-expression. However, her description of the 'natural' elements of the play park contrasted starkly with early accounts and images of bombsite playgrounds, with piles of debris, little greenery and the only infrastructure a tall wire fence.

Beyond the bombsite

While some adventure playgrounds were short-lived and they were never commonplace, both Allen and her values came to influence wider playground thought. This impact is often underemphasised in accounts of her work and in narratives associated with the adventure playground movement, perhaps because it is less iconic than images of children playing in rubble and because it took a little longer to have an effect. Allen later recalled that 'in the public mind, I was identified with adventure playgrounds. In fact, my interests had always been broader'.[110]

Perhaps as a result of her long-standing professional relationship with the landscape architect Richard Sudell, Allen became increasingly involved in the NPFA and its playground campaigning work. Like Allen, Richard Sudell (1892–1968) was a founding member of the Institute of Landscape Architects and for three years in the 1930s Allen and Sudell worked together, most notably on the roof garden at Selfridges.[111] In 1937, Sudell promoted Wicksteed Park as an exemplar children's playground, but by the 1950s he had moved away from providing manufactured equipment in the spaces he designed.[112] In the early 1950s, he prepared designs for St Chads Park and Central Park in Dagenham and included felled trees as climbing structures in place of a steel climbing frame.[113] As gardening editor for *Ideal Home* magazine he promoted modernist and child-friendly domestic garden design.[114] Sudell became involved in the NPFA in 1950, and from 1952 was a member of both its children's playground committee and technical subcommittee, which coordinated the publication of *Playgrounds for Blocks of Flats*.[115] By 1954, Allen was also a member of the children's playground committee but its March 1954 meeting was dominated by discussions about pin badges, Harrods sports week and a gala dinner fundraiser; the only mention of adventure playgrounds was to note that Bethnal Green borough council had decided not to open one.[116] Allen found such meetings highly conservative, with an atmosphere that was hierarchical and deferential, rather than experimental or dynamic, but the NPFA's organisational structures and resources did help Allen to raise awareness of alternative playground ideas.[117]

The initial conservatism of the NPFA also extended to the park profession. Junk playgrounds were rarely mentioned in trade journals in the late 1940s or early 1950s. Even in an account of Copenhagen's open spaces, written by the city's director of parks in 1948, the Emdrup playground did not receive a mention.[118] A 1955 editorial in the *Journal of Park Administration* seemingly idealised a nineteenth-century conception of the children's playground, where spaces were 'fenced, levelled and drained, with a semi-permanent dry surface and restricted to the use of infants'.[119] Adverts in the same issue hint at the enduring sensibilities of park administrators at this time. A metalwork company based in Thetford, IRS Ltd, promoted its finest enamel 'keep off the grass' and 'no cycling' signs, while Wicksteed & Co. and its equipment remained on the front cover. In a letter to *The Times* in 1957, Manchester's director of parks and cemeteries felt that 'old-fashioned swings are still the most popular type of playthings for children', while sandpits and adventure playgrounds were apparently both unpopular and dangerous.[120]

Allen's aspiration to introduce unkempt and creative spaces for play was often at odds with park superintendents' simultaneous efforts to keep both children and 'nature' under control. Reginald Wesley, director of parks and cemeteries in Belfast, was indicative of wider values when he emphasised the significant benefits associated with new chemical weedkillers, fungicides and pesticides, while at the same time complaining about the behaviour of children.[121] For A. Dodds, fellow of the Institute of Park Administration, the appearance of the adventure playground and its 'deplorable collection of rubbish' was a major obstacle to its wider uptake. Dodds suggested that a new title – the 'unorthodox play area' – combined with new building materials, rather than debris, and more hygienic surroundings would appeal more to the wider parks profession and the politicians who governed their work.[122] However, for park administrators, the orthodox children's playground remained an item of equipment that could be purchased from commercial suppliers. In trade journals during the 1950s and 1960s, adverts for play equipment were positioned next to those for other day-to-day necessities of the parks department, including mowers, glasshouses, wirework litter bins, teak seats, seeds and chemical pesticides. A similar pattern could be found at trade events and exhibitions.[123]

Some in the profession – and beyond – were starting to feel that parks administrators were not moving with the times. There were repeated calls in the trade press to give up on nineteenth-century conceptions of the park, focused on lavish horticultural displays, and to instead adopt new approaches to leisure.[124] Even the government's 1960 Albemarle inquiry into the provision of services for young people felt that 'park committees often work jointly with cemetery committees, and they become dedicated only too easily to the task of keeping people off or under the grass'.[125] Despite the apparent impenetrability of the profession to new playground ideas, Allen was influencing play space thinking in Britain and beyond.

The conservative tendencies of the park profession were at odds with increasing evidence that children were not using the playgrounds that had been provided for them. A sociological study of the Lansbury Estate in 1954 found that while early residents felt it was a good place to live, most children played in the streets rather than the playgrounds.[126] Research for *Playgrounds for Blocks of Flats* echoed these findings; during 104 visits to 96 sites, playgrounds were only being used by children on 44 occasions. For Allen, this meant that playgrounds needed to provide a greater variety of play opportunities, something that could be achieved through the provision of play leaders and features that were more flexible and creative.

As we have already seen, play leadership was discussed and promoted before Allen became involved in playground advocacy, but her emphasis on the role of the play leader in junk playgrounds helped to legitimise wider efforts to promote adult involvement in children's play. In 1956, the NPFA produced a film on play leadership and from the late 1950s there were play leadership schemes operating in many towns and cities, including Ramsgate, Belfast and London.[127] By 1965, there were sixty schemes operating across the country, the NPFA provided grants to cover play leader salaries and worked with the Institute of Park Administration to offer an annual play leadership summer school.[128] However, park-based play leadership was often socially conservative in the activities offered, echoing Krista Cowman's findings in the adventure playground movement. Folk dancing for girls and sport for boys echoed nineteenth-century efforts to promote rational recreation, rather than Allen's notion of child-centred play supported by inconspicuous adults. By 1970, a separate Institute of Playleadership was established and included Allen and other notable play workers and advocates, including Drummond Abernethy, Joe Benjamin and Donne Buck.[129] Despite these efforts, most playgrounds remained unsupervised.

Allen's calls for greater flexibility and creativity in playground provision, as well as her emphasis on providing more 'natural' play opportunities, were increasingly evident in contemporary playground discourse. In 1954, a public exhibition and conference on children's playgrounds demonstrated the shift towards more diversity in playground thought and form. The week-long Children's Playground Conference and Exhibition was organised by the London branch of the NPFA to promote the urgent need for more dedicated spaces for children 'on the grounds of health as well as keeping them out of danger and mischief'.[130] In doing so, it combined traditional ideas about the role of the playground as a site of safety, health and social good, with modern communication technology, international networks and a greater emphasis on public engagement. Opened by the duke of Edinburgh, the conference was free to enter, welcomed the public and included exhibits from over thirty local authorities, landscape architects and equipment manufacturers.[131] The event introduced the general public to existing and new notions of the playground and highlighted the wider range of professionals interested in the design and layout of play spaces for children.

A specially commissioned film, *Come out to Play*, sought to showcase the development of innovative ideas in play space design.[132] The film provides an insight into the ongoing problem of children's place in public space, as well as the increasing diversity in playground thinking. The film's opening sequence shows a police officer discouraging a group of children from playing in a park, hinting at the ongoing tension between public parks as communal spaces of health and recreation and the perceived problems of unsupervised children and their behaviour. Evicted from the park, the children are shown playing in the street, at risk from motor traffic and a threat to nearby private property, while the narrator emphasises the need for proper playgrounds close to every home. Having set the scene, the film moves on to tentatively highlight the latest ideas in playground design. It did not reject the orthodox playground out of hand and includes extensive footage of the US film star Betty Hutton opening a new orthodox playground on Bermondsey council's Arnold housing estate.[133] According to the film's narrator, at £580,000 (£7,250) it was more than usually expensive, while the images showed conventional playground equipment, including swings, slides and rocking horses. The film also included footage from Clydesdale Road adventure playground, showing children around a fire, using makeshift swings, playing war games and being organised by a play leader. Unlike Emdrup, with its purpose-built boundary, building and planting, the Clydesdale site seems to have been little adapted since it was cleared of bomb debris. A chain link fence and small wooden shed seem to be the main adaptations. The segment of the film that aligns most closely with Allen's wider vision for children's play is set in Holland Park in London. Parts of the wild and overgrown wood were designated as a space where children could climb, dig and make dens.

In addition to the film, the accompanying conference papers and exhibition spoke to the increasingly diverse interest in the form and function of children's play spaces. Nottingham's director of parks, W.G. Ayres, felt the need for playgrounds was primarily a road safety matter and he expressed doubt about experimental ideas in playground design.[134] Equipment manufacturers, including Hirst, Hunt, Spencer Heath and George, and Wicksteed promoted their projects. Magistrate and youth club advocate, Basil Henriques, emphasised the playground's role in reducing juvenile delinquency, while the director of the Royal Society for the Prevention of Accidents highlighted the ongoing dangers of the street.[135] In contrast, a number of speakers and exhibits emphasised alternative approaches to the playground and its form. Richard Sudell spoke on children's playgrounds in the modern landscape, while Marjory Allen discussed adventure playgrounds. The accompanying exhibition was designed by the LCC's architects under the supervision of chief architect Leslie Martin, noted designer of London's Royal Festival Hall. The exhibition included plans and photographs from the landscape architect Sylvia Crowe on her play-related work for Harlow New Town Development Corporation and photos of Emdrup from the Danish Embassy. The increasing role of professional designers in play space creation will be explored later, but here it is interesting to note the variety of play spaces on display.

To make sense of both existing and emerging ideas, the exhibition designers established and presented a playground typology, and in doing so attempted to make sense of contemporary playground discourse. The first category in their typology was equipped playgrounds. This type was further subdivided into 'orthodox' spaces with swings and slides to promote physical movement; 'feature' play spaces with sandpits, concrete boats and decommissioned steam rollers to inspire fantasy and make believe; and 'commando' playgrounds incorporating tree trunks, suspended tyres and concrete pipes to provide 'not only a free and varied outlet for energy but a spur to imagination and invention'. A second category was unequipped playgrounds, comprising a flat area for ball games. The third category was natural playgrounds with undulations, banks, trees and bushes as an environment for creative play. Adventure playgrounds were the fourth typology, a space where 'destruction and vandalism are transmuted into creative effort, team spirit is fostered and leaders emerge'. The fifth and final category was the traffic playground, which would simultaneously 'provide amusement and teach road safety'.[136] The typologies and their descriptions show how the playground was now expected to provide a wider range of benefits to children and society. Playgrounds would provide space for both physical exercise and cerebral creativity, an outlet for excess energy and site for semi-structured games, a space to both interact with nature and learn to cope with the hazards of the modern world. By providing these benefits, the antisocial child could be transformed into a well-rounded leader and team player. Playgrounds were not to be segregated by age or gender, while greater freedom in play was meant to be a feature of such spaces.

But these assumptions masked more conservative approaches to understanding the way that children should play. Normative assumptions about how girls and boys should play were clear in the way spaces were described and despite the rhetoric around freedom, girls were largely missing from these accounts of the ideal playground. Cowboys, supermen and other male heroes were the ideal characters who would be embodied in imaginary play, while ball games areas provided space to play male-dominated sports such as football and cricket. Girls were not explicitly excluded from these spaces, but the terminology used to frame them was heavily dependent on forms of play most strongly associated with boys, echoing the wider provision of outdoor recreational facilities, which supported sports that were played by and seen as appropriate for men. If implemented and used in the way imagined by the exhibition curators, the playground would reinforce and perpetuate an inequitable presence in public space for girls and boys.

The 1954 exhibition was not the first time that these ideas were expressed, but it was the first time that they were brought together in one place. It is not clear how many people attended the exhibition, nor how widely *Come out to Play* was distributed. Nevertheless, with the publication of the prosaically titled guidance note *Selection and Layout of Land for Playing Fields and Playgrounds* (1956), the NPFA brought these discussions to a wider audience.[137] Prepared by R.B. Gooch, the NPFA's technical advisor, it was reprinted several times over the next decade. Gooch welcomed the move away from the 'monotony' of playgrounds dominated by orthodox tubular steel equipment, something made possible in part because the booklet did not include nor rely on adverts from play equipment manufacturers. Instead, he echoed Allen's call for greater diversity in play provision, a sensitivity to children's expectations and opportunities to interact with nature.

The most notable break with earlier NPFA guidance was an apparent recognition that children should be given 'the opportunity of doing what they want to, rather than what grown-ups think they ought to do'.[138] However, this was still meant to take place in the playground, rather than in the wider urban environment. As a result, *Selection and Layout* proposed the ideal comprehensive playground as one which still provided space for physical movements such as swinging, sliding, jumping and climbing, but also room for creative activities, making things, imaginary games, playing with sand and water, and even less energetic pursuits such as reading or playing dominoes. It acknowledged that children had diverse personalities and interests, that child development relied on more than just steel swings and slides, and that the playground should help to meet children's creative and cognitive growth. It also marked a renewal of efforts to reintroduce nature into the playground and encouraged improvisation on the part of adult playground designers. A small, single-page sketch included in *Selection and Layout* was reproduced and distributed by the NPFA as a larger drawing. *Sketch Suggestions of Improvised Equipment for Children's Play* showed how more naturalistic materials such as trees, logs, grass mounds, sand and rocks could all help to make good places to play, while other materials and forms, such as concrete tunnels, brick walls and replica trains and boats could all promote imaginary play (Figure 4.2).[139] An added benefit was that this type of play space could potentially be created for little cost, using local materials and voluntary labour.

This ideal type would be restated in many of Marjory Allen's later publications, including *Design for Play*, *Play Parks* and *Planning for Play*, and in her evidence to the Parker Morris inquiry into housing standards.[140] Although best remembered for establishing domestic space requirements, the latter also made recommendations for play provision. In calling for sand, water, rough ground and tools, along with an emphasis on imaginative and creative play, it was clearly influenced by Allen's ideas. At the same time, by acknowledging that estate landscapes needed to accommodate both space for play and space for car parking, it highlighted wider tensions about how public space should be distributed and used. The problem of securing space for play in the face of urban redevelopment, increased car ownership and anxiety about juvenile delinquency was not confined to Britain, and an increasingly connected international network of play space campaigners, including Allen, shared ideas and experiences during the 1950s and beyond.

Allen's promotion of alternative visions for the playground in Britain coincided with her advocacy role with UNICEF in Europe and a wider renewed enthusiasm for international play networks. There had long been an exchange of ideas about dedicated public spaces for play, including links between British and US playground advocates from the 1890s, while the international diplomatic community had adopted the Geneva Declaration on the Rights of the Child in 1924. But after 1945 there was a

Figure 4.2 (right): Sketch Suggestions of Improvised Equipment for Children's Play by R.B. Gooch, National Playing Fields Association, 1956, London Metropolitan Archives, CLC/011/MS22287.

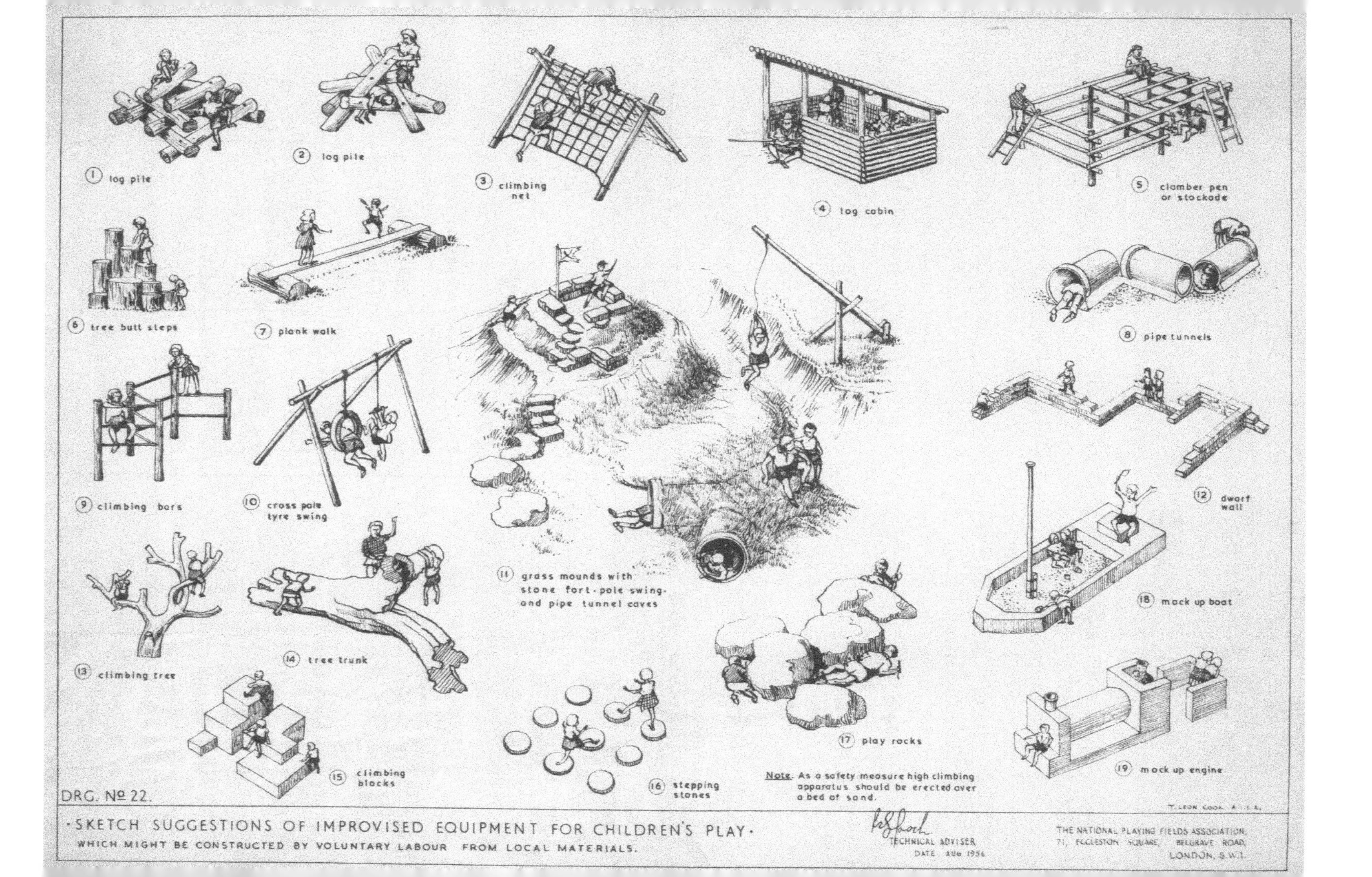
1 log pile
2 log pile
3 climbing net
4 log cabin
5 clamber pen or stockade
6 tree butt steps
7 plank walk
8 pipe tunnels
9 climbing bars
10 cross pole tyre swing
11 grass mounds with stone fort-pole swing-and pipe tunnel caves
12 dwarf wall
13 climbing tree
14 tree trunk
15 climbing blocks
16 stepping stones
17 play rocks
18 mock up boat
19 mock up engine
Note. As a safety measure high climbing apparatus should be erected over a bed of sand.
DRG. Nº 22.
T. LEON COOK A.I.L.A.
·SKETCH SUGGESTIONS OF IMPROVISED EQUIPMENT FOR CHILDREN'S PLAY·
WHICH MIGHT BE CONSTRUCTED BY VOLUNTARY LABOUR FROM LOCAL MATERIALS.
TECHNICAL ADVISER
DATE AUG 1956
THE NATIONAL PLAYING FIELDS ASSOCIATION,
71, ECCLESTON SQUARE, BELGRAVE ROAD,
LONDON, S.W.1.

significant increase in international cooperation. Landscape historian Jan Woudstra has suggested that Scandinavian countries in particular led a move away from equipment-dominated playgrounds towards a greater emphasis on nature, fantasy and personal creativity.[141] While Sweden and Denmark were often held up as exemplars in play space provision and design, the exchange of ideas and information took place far beyond northern Europe.

British trade journals included international case studies, exploring play spaces in Europe and the US, and park departments hosted overseas visitors.[142] The first International Conference in Park Administration took place in London in 1957 and included, alongside exhibition materials from Colwyn Bay, Copenhagen and China, a presentation on children's playgrounds by Allen and a trade exhibition that included Hunt and Wicksteed.[143] The conference led to the formation of the International Federation of Park Administrators (1957) and was followed by a United Nations seminar on playgrounds in 1958, a second world congress in 1962, attended by over a thousand delegates from twenty-six countries, and a third world congress in 1967.[144] One commentator concluded that 'when so much attractively designed playground equipment is being produced – especially on the Continent – it is somewhat melancholy to see new playgrounds in Britain fitted out with equipment that was probably designed around the turn of the century'.[145] Even in other countries playground equipment was not meeting the expectations of some. Arvid Bengtsson, the director of parks in Helsingborg, Sweden, felt that 'playground equipment which is on sale in this country is somewhat unimaginative and conservative. We in the Parks Office try therefore to design and construct the equipment which is needed'.[146]

In moving away from standardised manufactured equipment, Bengtsson was one example of a wider shift in international thought, perhaps best demonstrated by the publication of *Spielplatz und Gemeinschaftszentrum* (*Playgrounds and Recreation Spaces*, 1959) in Stuttgart and London.[147] In addition to examples from Britain and Germany, it included creative play space designs from the Netherlands, Switzerland, Denmark, Sweden, France and Italy, India and Japan, Brazil and the USA. In his introduction, the Swiss play space advocate Alfred Ledermann linked the need for dedicated children's play spaces to the problems of the modern city, including its impact on the nerves and health of urban inhabitants and the lack of wild space for children to play. Inspired by the Dutch cultural historian Johan Huizinga's book *Homo Ludens* (1938) and its emphasis on the central place of play in human culture, Ledermann argued that urban life needed more opportunities for playfulness, from the design of homes and gardens to open space on housing estates and in the public parks. Examples of progressive play space designs from around the world showed how town planners, designers and educationalists could work together to reclaim spaces for play in the city. Although Allen seems not to have been involved in preparing *Playgrounds and Recreation Spaces*, she would use some of the examples in her later publications and many of the case studies undoubtedly matched Allen's idea of the ideal play space.

In summary, from the late 1940s, Allen sought to rejuvenate the imagined playground so that it corresponded more closely to contemporary notions of childhood and provided opportunities for more naturalistic play. Although popularly associated with adventure playgrounds, she had a far wider influence on play space thinking. The rhetoric that emphasised the need for playgrounds endured and the critiques she expressed had largely been initiated and developed by others, often in the interwar years. However, in bringing them together and making them more widely and publicly accessible, she had a significant impact on visions of the ideal playground. She exposed the tension between orthodox playground design and evolving ideas about the developmental needs of children. Although often overlooked, providing more 'natural' play opportunities was also an important motivation for Allen. But just like other attempts to introduce elements of nature into the city, her ideas about a natural childhood and naturalistic play spaces were a product of her particular experiences and values, rooted in a rural nostalgia, horticultural training and practical work experience. Operating at a variety of scales, she contributed to local playground committees, campaigned nationally on play space provision and was connected with and contributed to international discourse through multinational conferences and networks. Through her campaigning, Allen challenged conventional playground thinking and encouraged experimentation in play space design, something that professional designers would progressively replicate in urban reconstruction schemes and on new housing estates.

Reimagining the playground: artists and architects

If postwar planners routinely designated space for children's play in modern urban environments, those tasked with imagining and designing the buildings and landscapes that gave form to such settings increasingly engaged with the detailed form of the playground and its contents. This was not new in 1945 and artists and designers had been involved in shaping modern versions of childhood during the first half of the twentieth century. From Charles Rennie Mackintosh's role in Glasgow's turn-of-the-century school building programme, through Bauhaus toys in Germany, to Tecton's Finsbury Health Centre, designers were addressing children's education, entertainment and health.[148] The playground did not escape this attention either. In 1934, the sculptor Isamu Noguchi imagined a radical play space for Central Park in New York, but his *Play Mountain* was never realised.[149] Instead, it was after the Second World War that the playground became even more firmly embedded into creative responses to the city. From late 1940s to the early 1960s, designers created infrastructure for the welfare state, including schools, hospitals and play spaces, influenced by utopian ideas, social planning and modernist aesthetics. In 1954, the Museum of Modern Art in New York ran a Playground Sculpture exhibition, which one critic described as a 'strange and wonderful world of colour and shapes'.[150] However, this did not mean that the principle of architectural experiment in the realms of play space design was widely accepted across the Atlantic in Britain. A 1957 article in *The Architect*, for instance, promoted the ideal play space as one that closely resembled the orthodox playground, where swings, slides and other apparatus predominated.[151] Another architectural commentator concurred, suggesting that 'British finances and the British temperament are vaguely against the planned playground, except in its most conventional form as a collection of swings and seesaws'.[152]

Despite mainstream support for conventional play equipment, there was increasing criticism of the orthodox playground from within the design profession. In an idiosyncratic conference paper in 1947, Clough Williams-Ellis welcomed the gradually improving provision of public spaces for children, but reacted against the use of 'frightful' railings which invariably surrounded them (he also found the 'shrubbery-pokery' of many parks distressing and most garden decoration 'debased and repulsive').[153] The architect Archie McNab found that play equipment manufacturers produced 'a range of products which on the whole is pretty dismal and unimaginative ... often more suited to a gymnasium than to helping small children to enjoy themselves'.[154] As well as summing up the previous seventy years of playground thought, he provided numerous examples of what he felt was more imaginative and creative but still industrially produced play equipment. In contrast, a small number of designers moved away from commercial play equipment to redefine play space forms in far more creative ways.

The historian Elain Harwood has emphasised the significant role played by the architects of postwar council housing schemes in designing the surrounding landscapes.[155] The large-scale redevelopment schemes made possible by comprehensive planning and wartime bomb damage meant that architects were presented with an almost blank canvas when designing new housing estates. Existing streets and buildings were often cleared entirely, and designers were tasked with creating new urban environments, within the site boundary at least, where homes, open spaces and playgrounds could be carefully integrated. One of the earliest and most notable postwar examples was the Churchill Gardens estate in Pimlico, Westminster. Designed and laid out by Philip Powell (1921–2003) and John Hidalgo Moya (1920–94), the estate provided a high-density mix of homes in blocks of different heights and was one of the first large-scale housing schemes after the war. In addition to the buildings and road layout, Powell and Moya also carefully planned the landscaping in between, including the provision of open spaces and the design of structures for play. However, this was not part of their initial commission. Instead, it was a personal decision to consider play provision in this way and Powell's particular sense of childlike fun is evident from his letters to the building contractor, seeking an old steamroller for one of the play spaces.[156] The associated play structures made use of materials that were similar to those used for the estate buildings, including brick and concrete, as well as more irreverent forms, such as a flying saucer, and some items of conventional play equipment (Figure 4.3).

This emphasis on the play value of architectural details and building materials was echoed within the LCC architects' department, considered one of the foremost architectural practices in the world at the time. Finding that children were more interested in the steps, slopes, seats and bollards of estate landscapes than the unsatisfactory and often actively dangerous specialist playground equipment, the department set out to design its own play structures.[157] Architects produced sketches of playhouses, dodge walls, bollard seats and wooden tents that could be created by the same building contractors who would build the new homes (Figure 4.4). The drawings were inserted into the department's design guidance in 1959 and several of these structures, along with a water tray and sandpit, were installed in four experimental play spaces, including on the Barnsbury estate in Islington and Woodberry Down estate in Hackney.[158]

Having produced bespoke play structures, architects turned to the wider estate landscape. Future phases of the Churchill Gardens scheme attempted to create an urban environment where 'children charge straight from indoors to play on the grass between the maisonettes, and their parents sit out in deck-chairs in the summer'.[159] This romantic image of the council estate, with children playing and parents freed from work and childcare responsibilities, might say more about the expectations of the

Figure 4.3: Children's playground, Churchill Gardens estate by J. Maltby, 1963, John Maltby/RIBA Collections, RIBA34960.

author than the realities of life on the estate, but it did represent a significant shift in thinking. Rather than enclose equipment within a designated playground, the whole estate environment needed to be considered when providing spaces for children to play.

One response was a logical extension of earlier attempts to segregate children and motor vehicles. But rather than encourage children into specific playgrounds, cars would be restricted to roads, while children had greater freedom within the estate. As Powell and Moya worked on Churchill Gardens, the sociologist Charles Madge argued for new urban environments where children could play more freely, but in the very different setting of low-rise, low-density Stevenage new town. Without motor traffic nearby, footpaths could become the 'natural patrolling ground for tricycles and other children's wheeled vehicles', while 'garden commons' provided space for sandpits and games.[160] Eleanor Mitchell, the designer of the Notting Hill adventure playground, also argued that play opportunities should be widely distributed in small quantities throughout the urban landscape, to create spaces for children to play or talk to friends while parents did their shopping.[161] There were practical experiments with this approach to play provision. In the new town of Basildon, sculptural

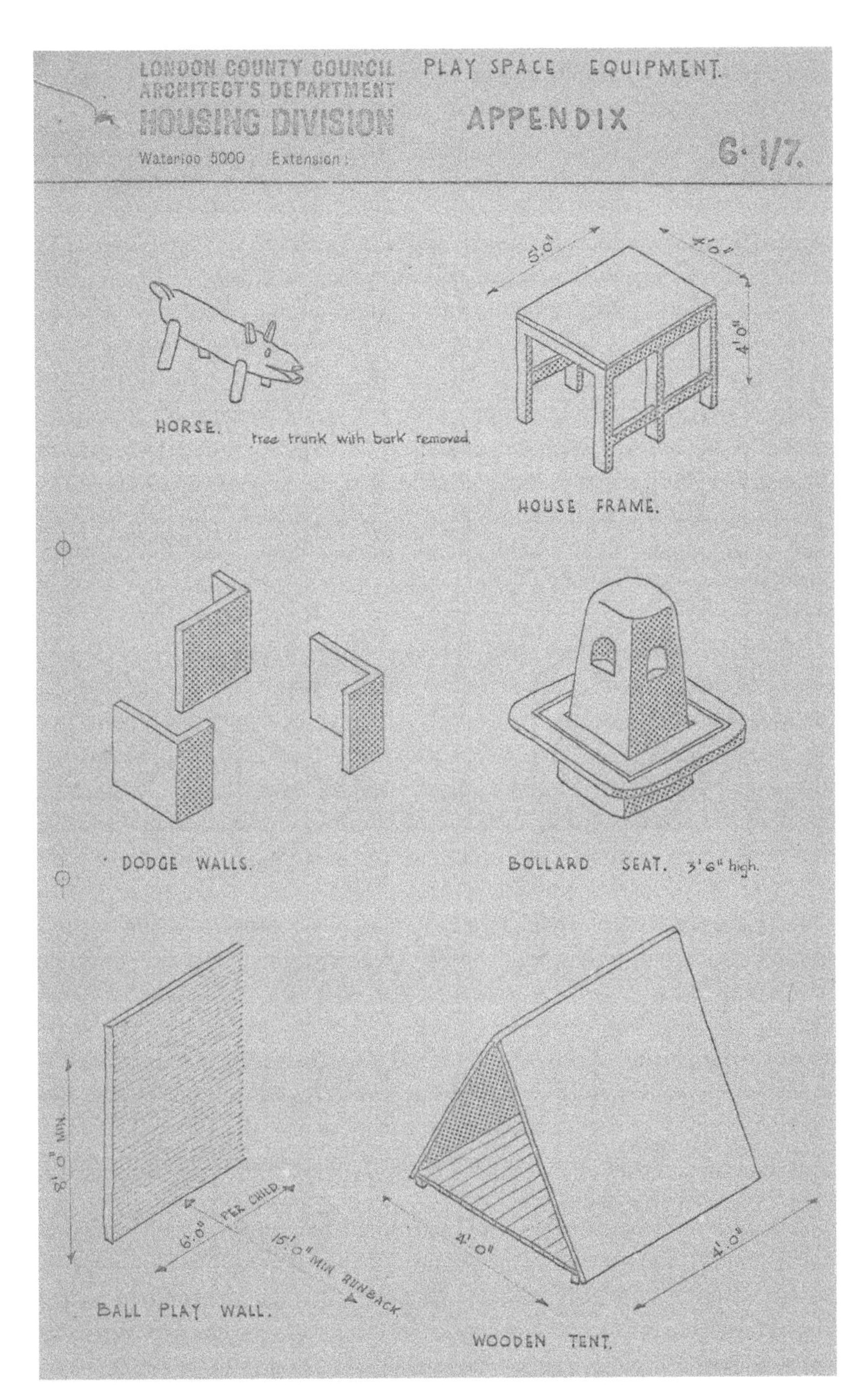

Figure 4.4: Experimental play equipment by LCC Architects Department, 1959, © London Metropolitan Archives (City of London), GLC/HG/HHM/12/S026A.

play equipment was scattered in car-free streets and squares, but when set in hard paved areas they seemed a long way from Madge's vision of a green garden common.

An alternative response came from the Netherlands, where architect Aldo van Eyck sought to create a more playful urban environment by reintegrating rather than segregating children from the city. Between 1947 and the 1970s, van Eyck created over 700 playable spaces in Amsterdam, mostly using bespoke sculptural installations that encouraged children to be creative and stimulated community life.[162] Invariably located close to homes but within the street setting, they tended to have little or no physical segregation from motor traffic.[163] Several of van Eyck's designs appeared in the 1959 English translation of *Playgrounds and Recreation Spaces*, but there seems to have been little wider acknowledgment of his radical approach among British play space advocates. The proximity of play to the perils of the street in van Eyck's schemes was too close for campaigners who had long emphasised that the street was not a place for play.

Instead, the involvement of architects in British play space design was most often associated with Brutalist housing estates in this period. In Sheffield, the city architect's department included concrete play structures on the Park Hill estate, while Erno Goldfinger produced sketches of alternative play forms and included Brutalist play spaces at the Balfron tower in Poplar, east London.[164] This form of experimentation – and in particular, the emphasis on architectural aesthetics rather than play function – was not welcomed by Marjory Allen. In fact, she was extremely critical of architectural involvement in play space provision. She argued that the orthodox playground, 'with fixed equipment chosen from an ironmonger's catalogue', represented one end of a dark spectrum and that at the other extreme were 'over-elaborate, over-clever, too slick' spaces designed by architects.[165] Neither swings and slides nor painted steamrollers and unalterable sculptural forms provided children with the freedom to play as they wished. In contrast, a few landscape designers were creating spaces that supported the free, creative and naturalistic play that Allen idealised.

A review of the contents of the Institute of Landscape Architects' journal from the 1930s to the 1970s found few contributions relating to the design of children's play space.[166] Despite this lack of coverage in the journal, there were landscape designers interested in play provision. As early as 1936, Thomas Adams had called for experts to be involved in shaping the modern city and specifically that landscape architects should be responsible for the creation of parks, playgrounds and promenades.[167]

After the war, it was the landscape architect Brenda Colvin (1897–1981) who most clearly elucidated a vision for children's play that combined Allen's explicit promotion of the potential of bombsites and her implicit appreciation of the natural environment and its benefits for children. In *Land and Landscapes* (1948), Colvin argued that designers should be promoting properly interconnected urban park systems, to bring fresh air and natural beauty within easy reach of all urban inhabitants. She shared Allen's assumptions about the playful needs of children, the playfulness of the rural landscape and the opportunities presented by the consequences of war. Colvin felt that children needed imaginative and adventurous play and that 'a good bomb crater, a tank trap, or a Home Guard dug out' all provided useful places were urban children could play.[168] She also suggested that these features represented an urban imitation of the play opportunities that were consistently available to children in the countryside. In an example from the open downs near Luton, she found children gathering to play on a steep chalk escarpment, with its gnarled tree roots, ropes and swings, mud slides, shrubs and trees. For Colvin, this environment provided freedom from the grown-up world and a haven for children's imagination. She suggested that when attempting to recreate similar play opportunities in the city, designers needed to provide irregularity, steep slopes, uneven ground, trees for climbing and swings, rough grass, water and surroundings that evoked a forest setting. Colvin worked on hundreds of schemes, from small gardens to industrial and institutional landscapes, but did not become known for creating children's play spaces. Instead, one of the most notable exponents of the naturalistic play spaces promoted by Colvin and Allen was Mary Mitchell.

Mary Mitchell (1923–88) qualified as a landscape architect in 1955 and briefly worked in Richard Sudell's practice and for the Stevenage New Town Development Corporation. But it was in her later work for Birmingham Corporation and in private practice that she established a reputation as a pioneering designer of children's playgrounds. Her work featured in a number of influential publications in Britain and overseas, including Marjory Allen's *Planning for Play* (1968) and Arvid Bengtsson's *Environmental Planning for Children's Play* (1970). Mitchell's designs were in stark contrast to the orthodox playground, with its levelled asphalt, metal fencing and standardised equipment, and instead were developed specifically for each site, making use of existing and new landscape features and responding to the character of the surrounding urban environment.[169] She felt that play areas needed to be imaginative and functional, active and sociable, creative and intimate, and free from pollution, all with a view to promoting frequent use by children.[170]

In Birmingham, her designs for the Kingshurst Hall estate, Pool Farm estate and Chamberlain Gardens play spaces incorporated mature and new trees to create a woodland-like setting, while undulating landforms included bespoke slides and climbing structures, and there were open, grassy areas for both active and imaginative play. On the Lyndhurst estate, a single row of granite setts embedded into the grass delineated only a nominal boundary between the play area and the wider estate landscape. In Nuneaton and Blackburn, Mitchell created spaces with similar characteristics, even if the individual designs were unique to each location. They included steep-sided slopes, water, trees, slides integrated into small hills and other bespoke play structures in a naturalistic setting. In the Lee Valley Regional Park in London, Mitchell combined the reclamation of an industrial landscape with new play opportunities, adapting a disused sewage works to create the Markfield Action Playground.[171]

As well as adapting the landscape to make it more playful, from 1959 Mitchell introduced sculpture to the playgrounds she designed. In particular, she worked with the artist John Bridgeman to create abstract, often animalistic forms in a number of Birmingham open spaces, including the Nechells Green Redevelopment Area and Hawkesley Farm Moat estate. It is interesting to note that even when experimental and creative approaches to the playground were implemented there was still a tendency towards standardisation and repetition, even if only on a small scale. The formwork for Bridgeman's concrete and brass slide sculpture at Nechells Green was designed to be reused at least four times.[172] Few of the sculptures now survive, although the installation at Curtis Gardens in Birmingham is now Grade-II listed. That Allen and others showcased Mitchell's play space designs in their books and publications is not surprising. With their organic aesthetic, landscaping and planting, diverse play opportunities and site-specific layouts, in many ways Mitchell's projects represented the ideal play spaces that Allen had long called for.

Another landscape architect who created play spaces that would receive wider acclaim at the time was Michael Brown (1923–96). From the mid-1960s, he designed a number of play areas in London, High Wycombe and Redditch, mainly on social housing estates. Although less naturalistic than Mitchell's play space designs, Brown used a simple palette of hard materials, often brick, to create incidental and durable opportunities for play. Brown felt that regular features of the urban landscape, such as steps, railings, walls and benches, were preferable to manufactured play equipment and that opportunities for creative and imaginative play should be a feature of all outside space.[173] At the Brunel estate in

Paddington, Brown created a monumental slide structure out of brick as part of his wider landscape scheme, a feature that was Grade-II listed by Historic England in 2020 (Figure 4.5). For Marjory Allen it was not the individual play structures that were his most notable achievement, but rather the approach to the wider estate grounds. Allen commended Brown's design for the Winstanley estate in Battersea because 'the entire landscape scheme has been conceived in terms of children's play activities', so that the 'total environment' was available for play.[174]

Figure 4.5: Brick slide on the Brunel estate, London, c.1974, Landscape Institute / Michael Brown, Museum of English Rural Life, AR BRO PH5/1/524B.

This approach to providing for children's play was not new in 1962 when Brown designed the Winstanley estate. In his wide-ranging review of public housing schemes in 1958, the noted architect A.W. Cleeve Barr concluded that 'inadequate facilities for children's play have constituted one of the most miserable features of British postwar housing schemes'.[175] And while he repeated many of the recommendations in other publications about the details of play provision, perhaps his most radical assertion was that the designers of a new housing estate needed to consider the total design of its communal environment. In many ways this was the antithesis of the playground. Rather than accept that the urban landscape was a hostile place for children and respond by providing dedicated places to play, these calls for total design represented a new way of thinking about the child in the city. Children had long experienced the wider city as a place to play, but now play advocates and designers were starting to appreciate that too. In 1965, the landscape architect Bill Gillespie concluded that 'we need to get away from this isolated idea of the parks towards an open space system fully integrated with the other elements of the city'.[176] Anarchists and urbanists, such as Colin Ward and Jane Jacobs, would develop this notion further in the late 1960s and 1970s, arguing that the functional segregation of the city, including the creation of dedicated places to play, not only failed to recognise the lived reality of urban life, but also contributed to the increasing hostility of the wider environment for children and adults alike. This shift from criticism of the playground form to condemnation of the entire playground principle is explored further in the next chapter.

From the 1940s to the 1960s, the children's playground provided a public space where social and environmental assumptions about childhood, child development, nature and the city could be played out and challenged. Long-standing rhetoric that was used to justify the need for playgrounds, including road danger, a lack of urban nature and protection against delinquency, remained central to continued efforts by town planners and play space campaigners to promote the need for dedicated play spaces for children. In numerical terms, the National Playing Fields Association distributed more financial support for municipal playground projects in the decade after the war than in the fifteen years before the war. At the same time, there was far greater experimentation in the playground form in response to new ideas about children's developmental needs, new forms of housing and the centrality of childhood to the new welfare state. Marjory Allen's promotion of the Emdrup junk playground was not a radical break with earlier thinking, even if it appeared very different to the traditional orthodox playground. Instead, Allen fused earlier critiques of the playground with changing constructions of childhood and her own conceptions of nature to spur high-profile and public discussion about the ideal playground. This in turn provided the critical space for advocates, designers and in a small number of cases children to experiment with play space form. This period of experimentation would be short-lived as commercial suppliers adapted their products and there were new concerns about playground safety. When combined with reductions in local authority funding and changing leisure habits in the later twentieth century, the playground ideal would face an existential threat.

Notes

1. Harry Hendrick, *Child Welfare, England 1872–1989* (London: Routledge, 1994); Mark Swenarton, Tom Avermaete and Dirk van den Heuvel, eds., *Architecture and the Welfare State* (London: Routledge, 2015).

2. Sir John Anderson to Lawrence Chubb, Letter, 25 April 1944, National Archives, CB 1/76.

3. National Playing Fields Association, 'Memorandum on the Recent Work of the Association', 1942, Museum of English Rural Life, SR CPRE C/1/73/1.

4. 'Bomb on East End School', *The Times*, 14 June 1917, p. 8; Niko Gärtner, 'Administering "Operation Pied Piper": How the London County Council Prepared for the Evacuation of Its Schoolchildren 1938–1939', *Journal of Educational Administration and History*, 42.1 (2010), 17–32.

5. 'The Childless City', *The Observer*, 10 September 1939, p. 11.

6. Peter Bruce Saunders, 'Oral History Interview, Reference to Playing in Damaged Buildings at 00:08:00', 1999, Imperial War Museum, Sound 18748; Leo Mellor, *Reading the Ruins: Modernism, Bombsites and British Culture* (Cambridge: Cambridge University Press, 2011).

7. Dudley S. Cowes, 'Leave This to Us, Sonny – You Ought to Be out of London', n.d., Imperial War Museum, Art.IWM PST 15093; 'Children in Towns', *The Times*, 11 March 1944, p. 5.

8. 'Shelter Play-Centres', *The Guardian*, 18 October 1941, p. 8.

9. F. Tennyson Jesse, 'Evacuation', *The Times*, 22 September 1939, p. 6.

10. 'Recreation for Evacuees', *The Guardian*, 18 December 1939, p. 10; 'Evacuated Children's Holidays', *The Times*, 27 July 1940, p. 7.

11. 'Are Our Child Exiles Happy? A Plea for Play in Reception Areas', *Journal of Park Administration, Horticulture and Recreation*, 4.9 (1940), 213–15.

12. 'Wicksteed: Our Works Are Now Fully Occupied on 100% War Work', *Journal of Park Administration, Horticulture and Recreation*, 5.11 (1941), front cover.

13. 'In Brief', *The Guardian*, 15 June 1940, p. 8; 'An Open Space', *The Guardian*, 26 April 1940, p. 4.

14. J. Lovatt and others, 'The Fattening of Pigs on Swill Alone: A Municipal Enterprise', *Empire Journal of Experimental Agriculture*, 11 (1943), 182–90; 'Paddington's Municipal Piggery: Two Years of Remarkable Progress', *Journal of Park Administration, Horticulture and Recreation*, 8.3 (1943), 35.

15. 'A Great Day for Paddington: Montgomery of Alamein Opens Rebuilt Recreation Ground', *Journal of Park Administration, Horticulture and Recreation*, 13.1 (1948), 29–30.

16. Alf T. Harrison, 'A Children's Paradise: The Children's Playground', *Journal of Park Administration, Horticulture and Recreation*, 11.7 (1946), 167–71.

17. B. Hirst & Sons to W.J. Hepburn, 'Ne Plus Ultra Playground Equipment Catalogue to Hyde Park Superintendent', 23 September 1949, National Archives, WORKS/16/391; Charles Wicksteed & Co., 'Price List for Playground Equipment', 1949, London Metropolitan Archive, GLC/RA/D2G/04/091.

18. 'Facilities Provided as a Result of Financial Assistance from the NPFA', *Playing Fields Journal*, 17.1 (1957), 12.

19. Marjorie Cruickshank, 'The Open-Air School Movement in English Education', *Paedagogica Historica*, 17.1 (1977), 62–74.

20. Lynn Cook, 'The 1944 Education Act and Outdoor Education: From Policy to Practice', *History of Education*, 28.2 (1999), 157–72.

21. Derek Fraser, *The Evolution of the British Welfare State* (London: Palgrave Macmillan, 2017).

22. F.J. Osborn, 'Pepler, Sir George Lionel (1882–1959)', *Oxford Dictionary of National Biography* (Oxford: Oxford University Press, 2004); George L. Pepler, 'Open Spaces', *The Town Planning Review*, 10.1 (1923), 11–24; National Playing Fields Association, *Third Annual Report* (National Playing Fields Association, 1929), National Archives, CB 4/1.

23. G.T. Eagleton, 'Wanted – a Standard for Small Playgrounds', *Playing Fields Journal*, 13.2 (1953), 48–50.

24. National Playing Fields Association, *Survey of Urban Playing Facilities* (London: National Playing Fields Association, 1951); 'Urban Areas Need More Games Facilities', *Playing Fields Journal*, 11.2 (1951), 32–5.

25. Local Government Board for Scotland, *Report of the Women's House-Planning Committee* (Edinburgh: HMSO, 1918); Ministry of Reconstruction Advisory Council, *Women's Housing Sub-Committee Final Report* (London: HMSO, 1919), National Archives, RECO 1/629.

26. John Boughton, *Municipal Dreams: The Rise and Fall of Council Housing* (London: Verso, 2019).

27. Simon Pepper and Peter Richmond, 'Upward or Outward? Politics, Planning and Council Flats, 1919–1939', *The Journal of Architecture*, 13.1 (2008), 53–90.

28. Mark Swenarton, 'Tudor Walters and Tudorbethan: Reassessing Britain's Inter-War Suburbs', *Planning Perspectives*, 17.3 (2002), 267–86.

29. Judith Roberts, 'The Gardens of Dunroamin: History and Cultural Values with Specific Reference to the Gardens of the Inter-war Semi', *International Journal of Heritage Studies*, 1.4 (1996), 229–37; Sophie Seifalian, 'Gardens of Metro-Land', *Garden History*, 39.2 (2011), 218–38.

30. Matthew Hollow, 'Suburban Ideals on England's Interwar Council Estates', *Garden History*, 39.2 (2011), 203–17.

31. Elizabeth Darling, '"The Star in the Profession She Invented for Herself": A Brief Biography of Elizabeth Denby, Housing Consultant', *Planning Perspectives*, 20.3 (2005), 271–300.

32. White City Estate Plan, LCC, London Housing (1937), p. 111 in Pepper and Richmond, 'Upward or Outward'.

33. 'Slum Clearance at Poplar', *The Times*, 30 August 1934, p. 6; 'Slum Clearance in London', *The Times*, 8 March 1938, p. 21.

34. J. Richardson, 'The Provision of Open Spaces in Slum Clearance Areas and Congested Districts', *Journal of Park Administration, Horticulture and Recreation*, 4.5 (1939), 125–9.

35. 'Quarry Hill Flats, Playground, Kitson House', 1939, Leeds Central Library, D LIE Quarry (12) https://www.leodis.net/viewimage/98774 [accessed 6 July 2023].

36. Matthew Whitfield, 'Keay, Sir Lancelot Herman (1883–1974)', *Oxford Dictionary of National Biography* (Oxford: Oxford University Press, 2004).

37. 'New Flats in Liverpool', *The Times*, 21 June 1935, p. 13.

38. Maxwell Fry, 'The New Britain Must Be Planned', *Picture Post*, 4 January 1941, 15–18.

39. Emmanuel Marmaras and Anthony Sutcliffe, 'Planning for Post-war London: The Three Independent Plans, 1942–3', *Planning Perspectives*, 9.4 (1994), 431–53;

Patrick Abercrombie and J.H. Forshaw, *County of London Plan* (London: Macmillan, 1943).

40. J. Paton Watson and Patrick Abercrombie, *A Plan for Plymouth* (Plymouth: Underhill, 1943), p. 2.

41. National Playing Fields Association, 'Minutes of the Ad Hoc Committee to Enquire into the Provision of Play Space on New Housing Estates, 17 May', 1960, National Archives, CB 1/64.

42. London County Council, 'Spa Fields Extension', 1951, London Metropolitan Archives, GLC/RA/D2G/04/091; 'Breathing Space in London', *Park Administration, Horticulture and Recreation*, 24.12 (1960), 667–9.

43. Harriet Atkinson, *The Festival of Britain: A Land and Its People* (London: Tauris, 2012).

44. 'Forward to Festival of Britain', *The Architectural Review*, 110 (1951), 73–9.

45. Lord Luke, 'Festival of Britain 1951 NPFA to Play Important Part', *Playing Fields Journal*, 9.4 (1949), 202.

46. Becky Conekin, *'The Autobiography of a Nation': The 1951 Festival of Britain* (Manchester: Manchester University Press, 2003).

47. Plan of Lansbury 'Live Architecture' Exhibition in Atkinson, *The Festival of Britain*.

48. Matthew Hollow, 'Utopian Urges: Visions for Reconstruction in Britain, 1940–1950', *Planning Perspectives*, 27.4 (2012), 569–85.

49. Simon Gunn, 'The Rise and Fall of British Urban Modernism', *Journal of British Studies*, 49.4 (2010), 849–69.

50. Anthony Minoprio, 'Crawley New Town', *Town and Country Planning*, 16.64 (1949), 215–21.

51. Watson Garbutt, 'A Village Becomes a New Town', *Town and Country Planning*, 12.45 (1944), 22–5.

52. Ministry of Health, *Housing Manual* (London: HMSO, 1949), p. 17.

53. National Playing Fields Association, 'Minutes of the First Meeting of the Children's Playground Technical Subcommittee on 4 November', 1952, National Archives, CB 1/68.

54. National Playing Fields Association, *Playgrounds for Blocks of Flats* (London: National Playing Fields Association, 1953), Museum of English Rural Life, P2870 Box 5/12.

55. Malcolm Tozer, 'A History of Eton Fives', *The International Journal of the History of Sport*, 30.2 (2013), 187–9.

56. National Playing Fields Association, *Playgrounds for Blocks of Flats*, p. 8.

57. Ministry of Housing and Local Government, 'Housing Handbook', 1957, p. 62, National Archives, HLG 31/11.

58. National Playing Fields Association, 'Minutes of the Ad Hoc Committee to Enquire into the Provision of Play Space on New Housing Estates, 17 May', p. 2; National Playing Fields Association, *Playgrounds for Blocks of Flats*, 6th edn (London: National Playing Fields Association, 1974), National Archives, CB 4/76.

59. Marjory Allen and Mary Nicholson, *Memoirs of an Uneducated Lady: Lady Allen of Hurtwood* (London: Thames and Hudson, 1975).

60. Bob Holman, *Champions for Children: The Lives of Modern Child Care Pioneers* (Bristol: Policy Press, 2001).

61. Hal Moggridge, 'Allen, Marjory, Lady Allen of Hurtwood (1897–1976), Landscape Architect and Promoter of Child Welfare', *Oxford Dictionary of National Biography* (Oxford: Oxford University Press, 2007).

62. Marjory Allen, 'Juvenile Delinquency', *The Times*, 6 December 1948, p. 5.

63. Elizabeth Denby, *Europe Re-Housed* (London: Allen & Unwin, 1938), p. 269.

64. Marjory Allen, 'Letter to the Editor: Children's Playgrounds', *The Times*, 12 December 1952, p. 9.

65. Marjory Allen, 'Children in "Homes"', *The Times*, 15 July 1944, p. 5; Marjory Allen, *Whose Children?* (London: Simpkin Marshall, 1945).

66. Gordon Lynch, 'Pathways to the 1946 Curtis Report and the Post-War Reconstruction of Children's Out-of-Home Care', *Contemporary British History*, 34 (2020), 22–43 (p. 28).

67. Holman, *Champions for Children*; Roy Kozlovsky, 'Adventure Playgrounds and Postwar Reconstruction', in *Designing Modern Childhoods: History, Space, and the Material Culture of Children; An International Reader*, ed. Marta Gutman and Ning de Coninck-Smith (New Jersey: Rutgers University Press, 2007), pp. 171–90.

68. Allen and Nicholson, *Memoirs of an Uneducated Lady*, p. 230.

69. Carol Dyhouse, 'The British Federation of University Women and the Status of Women in Universities, 1907–1939', *Women's History Review*, 4.4 (1995), 465–85.

70. 'ASLA Notes', *Landscape Architecture*, 21.2 (1931), 139–45; Allen and Nicholson, *Memoirs of an Uneducated Lady*, p. 98; Harriet Jordan, 'Mawson, Thomas Hayton (1861–1933), Landscape Architect', *Oxford Dictionary of National Biography* (Oxford: Oxford University Press, 2010).

71. Allen and Nicholson, *Memoirs of an Uneducated Lady*, pp. 30–31.

72. Marjory Allen, *New Houses, New Schools, New Citizens* (London: Nursery School Association of Great Britain, 1934), p. 5.

73. Marjory Allen, 'Why Not Use Our Bomb Sites like This?', *Picture Post*, 16 November 1946, 26–9.

74. Allen and Nicholson, *Memoirs of an Uneducated Lady*, p. 196.

75. Jan Woudstra, 'Danish Landscape Design in the Modern Era (1920–1970)', *Garden History*, 23.2 (1995), 222–41.

76. Ning de Coninck-Smith, 'Where Should Children Play? City Planning Seen from Knee-Height: Copenhagen 1870 to 1920', *Children's Environments Quarterly*, 7.4 (1990), 54–61.

77. A.K. Winship, 'Editorial – Playing in Sand', *The Journal of Education*, 70.16 (1909), 436.

78. Ning de Coninck-Smith, *Natural Play in Natural Surroundings: Urban Childhood and Playground Planning in Denmark, c.1930–1950*, Working Papers in Child and Youth Culture, 6 (Odense: University of Southern Denmark, 1999).

79. G. Stanley Hall, *Youth: Its Education, Regimen, and Hygiene* (New York: Appleton, 1906); Essi Jouhki, 'Politics in Play: The Playground Movement as a Socio-Political Issue in Early Twentieth-Century Finland', *Paedagogica Historica*, 2023, 1–21.

80. Asbjørn Jessen and Anne Tietjen, 'Assembling Welfare Landscapes of Social Housing: Lessons from Denmark', *Landscape Research*, 46.4 (2021) https://doi.org/10.1080/01426397.2020.1808954; Anthea Holme and Peter Massie, *Children's Play: A Study of Needs and Opportunities* (London: Michael Joseph, 1970), p. 47.

81. Carl Theodore Sørensen, *Parkpolitik i Sogn Og Købstad* (Copenhagen: Christian Ejlers Forlag, 1931); quoted in Sven-Ingvar Andersson and Steen Høyer, *C. Th. Sørensen, Landscape Modernist*, trans. Anne Whiston Spirn (Copenhagen: Danish Architectural Press, 2001).

82. Sørensen, *Parkpolitik*, p. 54; quoted in Peter Bosselmann, 'Landscape Architecture as Art: C. Th. Sørensen. A Humanist', *Landscape Journal*, 17.1 (1998), 62–9 (p. 65).

83. Deborah Philips, *Fairground Attractions: A Genealogy of the Pleasure Ground* (London: Bloomsbury, 2012).

84. Robin Whalley and Peter Worden, 'Forgotten Firm: A Short Chronological Account of Mitchell and Kenyon, Cinematographers', *Film History*, 10.1 (1998), 35–51.

85. Philip J. Deloria, *Playing Indian* (New Haven: Yale University Press, 1998).

86. Scott Johnston, 'Courting Public Favour: The Boy Scout Movement and the Accident of Internationalism, 1907–29', *Historical Research*, 88.241 (2015), 508–29.

87. Joyce Antler, 'Mitchell, Lucy Sprague (1878–1967), Educator', *American National Biography* (Oxford: Oxford University Press, 1999).

88. Jean Lee Hunt, *A Catalogue of Play Equipment* (New York: Bureau of Educational Experiments, 1918), Project Gutenberg, 28466 www.gutenberg.org/ebooks/28466 [accessed 6 July 2023].

89. Patrick Geddes, *Cities in Evolution: An Introduction to the Town Planning Movement and to the Study of Civics* (London: Williams & Norgate, 1915), p. 97.

90. Ministry of Reconstruction Advisory Council, *Women's Housing Sub-Committee Final Report*, p. 11.

91. Margaret McMillan, *The Nursery School* (London: J. M. Dent & Sons, 1921).

92. Allen and Nicholson, *Memoirs of an Uneducated Lady*, pp. 157–9.

93. 'The Job of the Games Warden', *Playing Fields Journal*, 12.1 (1952), 25–6.

94. Krista Cowman, '"The Atmosphere Is Permissive and Free": The Gendering of Activism in the British Adventure Playgrounds Movement, ca. 1948–70', *Journal of Social History*, 53.1 (2019), 218–41.

95. Edward Keeling, *House of Commons Debate, 30 March 1944, Vol.398, Col.1542* (Hansard, 1944).

96. Ben Highmore, 'Playgrounds and Bombsites: Postwar Britain's Ruined Landscapes', *Cultural Politics*, 9 (2013), 323–36; Krista Cowman, 'Open Spaces Didn't Pay Rates: Appropriating Urban Space for Children in England after WW2', in *Städtische Öffentliche Räume: Planungen, Aneignungen, Aufstände 1945–2015 (Urban Public Spaces: Planning, Appropriation, Rebellions 1945–2015)*, ed. Christoph Bernhardt (Stuttgart: Franz Steiner Verlag, 2016), pp. 119–40.

97. Kozlovsky, 'Adventure Playgrounds and Postwar Reconstruction'; Highmore, 'Playgrounds and Bombsites'; Cowman, 'Open Spaces Didn't Pay Rates'; Lucie Glasheen, 'Bombsites, Adventure Playgrounds and the Reconstruction of London: Playing with Urban Space in Hue and Cry', *The London Journal*, 44.1 (2019), 54–74.

98. Marcus Lipton, *House of Commons Debate, 13 March 1953 Vol.512 Col.1735* (Hansard, 1953).

99. Institute of Park Administration, *Report of the 1955 Annual Conference* (London: Journal of Park Administration Ltd, 1955), Museum of English Rural Life, P2870 Box 5/39.

100. National Playing Fields Association, 'Minutes of the Children's Playground Committee on 17 March', 1954, National Archives, CB 1/59.

101. Marjory Allen, *Adventure Playgrounds* (London: National Playing Fields Association, 1953); National Playing Fields Association, 'Steering Committee on Adventure Playgrounds', 1953, National Archives, CB 1/53; National Playing Fields Association, 'Report of Adventure Playground Conference', 1956, National Archives, CB 1/67.

102. National Playing Fields Association, 'Minutes of the Children's Playground Committee on 18 November', 1954, p. 2, National Archives, CB 1/59.

103. 'Liverpool's Adventure Playgrounds', *Park Administration, Horticulture and Recreation*, 24.10 (1960), 515–17.

104. Allen and Nicholson, *Memoirs of an Uneducated Lady*, p. 237.

105. Rose Macaulay, *Pleasure of Ruins* (London: Weidenfeld & Nicolson, 1953).

106. Gabriel Moshenska, 'Children in Ruins', in *Ruin Memories: Materiality, Aesthetics and the Archaeology of the Recent Past*, ed. Bjørnar Olsen and Þóra Pétursdóttir (Abingdon: Routledge, 2014), pp. 230–49.

107. Institute of Park Administration, *Report of the First International Congress in Public Park Administration* (London: Journal of Park Administration, 1957), p. 67, RHS Lindley, 969.2 Ins.

108. Marjory Allen, 'Why the Stockholm Playgrounds Are So Successful', *The Architect and Building News*, 30 December 1954, pp. 812–14.

109. Marjory Allen, *Play Parks*, 3rd edn (London: Housing Centre, 1964), p. 10.

110. Allen and Nicholson, *Memoirs of an Uneducated Lady*, p. 245.

111. Annabel Downs, 'Sudell, Richard (1892–1968), Landscape Architect and Author', *Oxford Dictionary of National Biography* (Oxford: Oxford University Press, 2009).

112. Richard Sudell, 'How Can We Make Our Parks Brighter?', *Municipal Journal and Public Works Engineer*, 46.2300 (1937), 397–8.

113. Richard Sudell, 'Park Design for Modern Needs', *Journal of Park Administration, Horticulture and Recreation*, 14.10 (1950), 343–6; Richard Sudell, 'Wanted – More Play Leaders', *Playing Fields Journal*, 12.1 (1952), 20–24.

114. Seifalian, 'Gardens of Metro-Land'.

115. National Children's Playground Association, 'The Five Million Club, Minutes of Executive Committee, 17 October', 1950, National Archives, CB 1/76; National Playing Fields Association, 'Minutes of the First Meeting of the Children's Playground Technical Subcommittee on 4 November'.

116. National Playing Fields Association, 'Minutes of the Children's Playground Committee on 17 March'.

117. Allen and Nicholson, *Memoirs of an Uneducated Lady*, p. 235.

118. J. Bergmann, 'The Parks of Copenhagen', *Journal of Park Administration, Horticulture and Recreation*, 13.1 (1948), 18–21.

119. 'Editorial', *Journal of Park Administration, Horticulture and Recreation*, 20.7 (1955), 283.

120. R.C. McMillan, 'Changes in the Playground', *The Times*, 12 July 1957, p. 11.

121. Reginald Wesley, 'Play Leadership in the City of Belfast', *Park Administration, Horticulture and Recreation*, 24.8 (1960), 378–80.

122. A. Dodds, 'Play and Our Young People', *Park Administration*, 28.5 (1963), 36–43.

123. 'Ninth Exhibition of Park Equipment, Machinery and Materials', *Park Administration, Horticulture and Recreation*, 26.11 (1961), 46–8.

124. H.F. Clark, 'A New Type of Park Administrator', *Park Administration, Horticulture and Recreation*, 23.11 (1959), 439–41; 'Editorial', *Park Administration, Horticulture and Recreation*, 24.5 (1959), 199; 'Editorial', *Park Administration, Horticulture and Recreation*, 27.11 (1962), 21.

125. Ministry of Education, *The Youth Service in England and Wales* (London: HMSO, 1960), p. 68.

126. John Westergaard and Ruth Glass, 'A Profile of Lansbury', *The Town Planning Review*, 25.1 (1954), 33–58.

127. 'Play Leadership Film', *Playing Fields Journal*, 16.4 (1956), 25; J.H. Hingston, 'Play Leadership in the Borough of Ramsgate, Kent', *Park Administration, Horticulture and Recreation*, 24.9 (1960), 438–40; 'Come out to Play in Battersea Park', *Park Administration, Horticulture and Recreation*, 27.5 (1962), 65.

128. W.D. Abernethy, 'What Play Leadership Implies', *Park Administration*, 30.3 (1965), 55.

129. National Playing Fields Association, 'Institute of Playleadership Minutes, 9 February', 1970, National Archives, CB 1/64.

130. London and Greater London Playing Fields Association, 'Children's Playgrounds Exhibition and Conference Press Release', 1954, p. 1, London Metropolitan Archives, LCC/CL/PK/01/038.

131. 'Planning Children's Playgrounds', *The Times*, 12 June 1954, p. 3.

132. *Come out to Play*, 1954, British Pathé Archive, DOCS 1359.01 https://www.britishpathe.com/video/come-out-to-play-reel-1-1/ [accessed 6 July 2023].

133. 'Bermondsey Children's "Variety" Playground', *Playing Fields Journal*, 12.1 (1952), 35–7.

134. W.G. Ayres, 'The Provision of Children's Playgrounds by a Local Authority', *Journal of Park Administration, Horticulture and Recreation*, 19.4 (1954), 151–60.

135. 'New Playfield Ideas from All over the World: Fascinating Facts in Report on Children's Playground Exhibition', *Playing Fields Journal*, 15.1 (1955), 53.

136. London and Greater London Playing Fields Association, 'Children's Playgrounds Exhibition'.

137. R.B. Gooch, *Selection and Layout of Land for Playing Fields and Playgrounds* (London: National Playing Fields Association, 1956), National Archives, CB 4/59.

138. Gooch, *Selection and Layout of Land for Playing Fields and Playgrounds*, p. 66.

139. R.B. Gooch, *Sketch Suggestions of Improvised Equipment for Children's Play* (London: National Playing Fields Association, 1956), London Metropolitan Archive, CLC/011/MS22287.

140. Marjory Allen, *Design for Play: The Youngest Children* (London: Housing Centre, 1962); Allen, *Play Parks*; Marjory Allen, *Planning for Play* (London: Thames and Hudson, 1968); Ministry of Housing and Local Government, *Homes for Today and Tomorrow* (London: HMSO, 1961).

141. Jan Woudstra, 'Detailing and Materials of Outdoor Space: The Scandinavian Example', in *Relating Architecture to Landscape*, ed. Jan Birksted (London: E & FN Spon, 1999), pp. 53–68.

142. A.H. Garnsey, 'Playgrounds in Europe and America', *Playing Fields Journal*, 12.3 (1952), 33–5.

143. Institute of Park Administration, *Report of the First International Congress in Public Park Administration*.

144. International Federation of Park Administrators, 'Bulletin', 1967, Museum of English Rural Life, SR CPRE C/1/130/2; W. D. Abernethy, 'Report on United Nations Playground Seminar', 1958, National Archives, National Playing Fields Association CB 1/70; Institute of Park Administration, *Report of the Second World Congress in Public Park Administration* (London: Journal of Park Administration, 1962), RHS Lindley, 969.2 Ins; Institute of Park Administration, 'Third World Congress in Public Park and Recreation Administration Bulletin No.2', 1967, Museum of English Rural Life, SR CPRE C/1/130/2.

145. 'How Austria Equips Its Children's Playgrounds', *Park Administration, Horticulture and Recreation*, 22.9 (1958), 448–9 (p. 448).

146. Arvid Bengtsson, 'Children's Playground in a Swedish Town', *Park Administration, Horticulture and Recreation*, 22.10 (1958), 478–9 (p. 479).

147. Alfred Ledermann and Alfred Trachsel, *Playgrounds and Recreation Spaces*, trans. Ernst Priefert (London: The Architectural Press, 1959).

148. Juliet Kinchin and Aidan O'Connor, *Century of the Child: Growing by Design 1900–2000* (New York: The Museum of Modern Art, 2012).

149. Shaina D. Larrivee, 'Playscapes: Isamu Noguchi's Designs for Play', *Public Art Dialogue*, 1.1 (2011), 53–80.

150. Aline B. Saarinen, 'Playground: Function and Art', *New York Times*, 4 July 1954, p. 4.

151. 'Planning', *The Architect and Building News*, 1957, 477–81.

152. 'The Library Shelf', *Official Architecture and Planning*, 23.3 (1960), 135.

153. Clough Williams-Ellis, 'Biased Opinions', *Journal of Park Administration, Horticulture and Recreation*, 12.1 (1947), 15–17.

154. Archie McNab, 'Equipping Children's Playgrounds', *Design*, 159 (1962), 64–8.

155. Elain Harwood, 'Post-War Landscape and Public Housing', *Garden History*, 28.1 (2000), 102–16.

156. Elain Harwood, 'Review: The New Brutalist Image 1949–55 and The Brutalist Playground', *Journal of the Society of Architectural Historians*, 75.1 (2016), 117–19.

157. London County Council, 'Unsupervised Play Space on Housing Estates', 1959, London Metropolitan Archives, GLC/HG/HHM/12/S026A.

158. London County Council, 'Experimental Play Equipment: Sketch Layouts', 1959, London Metropolitan Archives, GLC/HG/HHM/12/S026A.

159. Philip Aldis, 'Churchill Gardens', *New Left Review*, 10, 1961, 55–9 (p. 58).

160. Charles Madge, 'Planning for People', *The Town Planning Review*, 21.2 (1950), 131–44 (p. 140).

161. Eleanor Mitchell, 'Planning for Children's Play', *Town and Country Planning*, 34.8–9 (1966), 418–21.

162. Rob Withagen and Simone R. Caljouw, 'Aldo van Eyck's Playgrounds: Aesthetics, Affordances, and Creativity', *Frontiers in Psychology*, 8.1130 (2017) https://doi.org/10.3389/fpsyg.2017.01130.

163. Liane Lefaivre and Ingeborg de Roode, eds., *Aldo van Eyck: The Playgrounds and the City* (Amsterdam: Stedelijk Museum, 2002).

164. Sam Lambert, 'Children Playing on the Climbing Frames in the Playground, Park Hill Estate, Sheffield', 1963, RIBA Collections, AP Box 212 Sheffield; Erno Goldfinger, 'Design for an Unidentified Playground', 1965, RIBA Collections, PA646/4(6).

165. Allen, *Planning for Play*, p. 18.

166. Ian C. Laurie, 'Public Parks and Spaces', in *Fifty Years of Landscape Design 1934–84*, ed. Sheila Harvey and Stephen Rettig (London: The Landscape Press, 1985), pp. 63–78 (p. 73).

167. Thomas Adams, *Outline of Town and City Planning: A Review of Past Efforts and Modern Aims* (London: J & A Churchill, 1936), p. 334.

168. Brenda Colvin, *Land and Landscape* (London: John Murray, 1948), p. 206.

169. Mary Mitchell, 'Birmingham Parks', *Park Administration*, 28.5 (1963), 47.

170. Mary Mitchell, 'Landscaping of Housing Areas', *Official Architecture and Planning*, 25.4 (1962), 193–6.

171. Mary Mitchell, 'Birmingham Playgrounds', *Playing Fields*, 21.4 (1961), 40–41; Mary Mitchell, 'Birmingham Playgrounds', *Playing Fields*, 24.2 (1964), 29–30; Mary Mitchell, 'Landscaping of Housing Areas', p. 193; 'An Imaginative Approach to Playground Provision', *Park Administration*, 28.12 (1963), 42; Allen, *Planning for Play*, p. 96; Arvid Bengtsson, *Environmental Planning for Children's Play* (London: Lockwood, 1970), pp. 104 & 148; 'Children's Playgrounds', *L'Architecture d'Aujourd'hui*, 165 (1973), xxxv; Tom Turner and Simon Rendel, *London Landscape Guide* (Dartford: Landscape Institute, 1983).

172. 'Playgrounds in Birmingham', *The Architect and Building News*, 218.24 (1960), 767–8.

173. Michael Brown, 'Landscape and Housing', *Official Architecture and Planning*, 30.6 (1967), 791–9; Michael Brown, 'Drawings and Plans in the Michael Brown Collection', 1966, Museum of English Rural Life, AR BRO DO.

174. Allen, *Planning for Play*, p. 26; Bengtsson, *Environmental Planning for Children's Play*, p. 52; Luca Csepely-Knorr and Amber Roberts, 'Towards a "Total Environment" for Children: Michael Brown's Landscapes for Play', in *Landscape and Children* (presented at the FOLAR Annual Symposium, Museum of English Rural Life, 2019).

175. A.W. Cleeve Barr, *Public Authority Housing* (London: Batsford, 1958), p. 46.

176. William Gillespie, 'Landscaping Our Urban Areas', *Park Administration*, 30.11 (1965), 40–43 (p. 43).

Chapter 5

Playground scuffles: 'it's ours whatever they say'

From the early 1970s, high-profile discussion about the ideal playground was amplified by wider debate in relation to childhood and cities, politics and economics. Over the next four decades, an enduring belief in the power of play was gradually challenged by shifting conceptions of the city, anxiety about children's safety and the changing status of local government and urban planning. Often characterised by contemporary politicians and the media as a decade beset with crisis, more recent revisionist accounts of the 1970s have contended that while Britain undoubtedly experienced a convulsive moment, the talk of crisis is significantly overstated. Such accounts instead contend that the decade is best characterised by a 'battle of ideas' in the media, publishing, higher education and politics.[1] The notion of a battle of ideas is also a useful characterisation of playground discourse from the late 1960s through to the early twenty-first century. As such, this chapter seeks to extend revisionist accounts of the era, pointing to an ongoing fermenting of thought, which started in earnest in the late 1940s and continued into the 1970s and beyond. It highlights the continuing place of the playground in visions of a modern, planned and healthy urban environment, before moving on to explore in more depth the contested place of the playground in local politics, national policy, sociological research and anarchic thought. It charts the increasingly polarised attitudes to play provision, identifying areas of conflict between radical play work and more traditional notions of the playground, as well as a subsequent widespread preoccupation with safety that significantly affected both approaches. Campaigners' concern about the danger of motor vehicles was largely superseded by wider anxiety about the threat posed to children by paedophiles, pets and, increasingly, playgrounds themselves.

In his influential book, *Lost Freedom*, historian Mathew Thomson has sought to make sense of efforts in the 1970s to promote greater freedom in urban childhood.[2] Compared to the present, the decade appears to be a time of considerable freedom for children, particularly their ability to experience the outside world and play in the city without parental supervision. At the same time, the decade was marked by calls for greater freedoms for the urban child, particularly from radical progressives. In attempting to make sense of this apparent paradox, Thomson argues that anxiety about the impact of the Second World War on children, combined with postwar concern about the danger from traffic, resulted in efforts to protect young people from the dangers of the city, including the creation of playgrounds. As we have already seen in earlier chapters, there is a much longer history to the creation of dedicated play spaces as a route to safety, health and happiness, particularly as a response to the dangers of traffic. The extended chronology examined here does not discredit Thomson's argument, but rather lends weight to his assertion that by the early 1970s the foundations for a reaction against the over-protection of children were well established.

This might seem at odds with the emphasis on childhood freedom explored in earlier chapters, particularly among those inspired by Marjory Allen's campaigning from the 1940s. Making sense of this requires some thought about how the term 'freedom' is being used. For Thomson and radical campaigners in the 1970s, it represented ideas about children's ability to play in and move through the urban environment, the distance they could travel from home, and a lack of direct adult supervision. However, for adventure playground advocates of the late 1940s and 1950s, the concept of freedom had primarily related to an individual child's ability to play in an instinctive and unstructured way, without the constraints imposed by standardised manufactured play equipment, asphalt and fencing. For advocates such as Allen, the need for dedicated places to play remained convincing and the form of the playground and the type of play that it facilitated were central to their ideas and actions. They imagined that childhood independence operated within the boundaries of the playground, while later radicals promoted autonomy for children on a city-wide scale. In Thomson's analysis, calls for greater childhood freedom in the 1970s were partially inspired by the limits on childhood play and mobility that the principle of the playground imposed, but also the anarchic possibilities that childhood independence and self-determination invoked.

This chapter considers how the principle of the children's playground was positioned in these debates about childhood freedom from the late

1960s. It examines how a postwar focus on the play of the individual child expanded to incorporate a wider political mission to reclaim the city for children, how those responsible for promoting and managing playgrounds reacted and the extent to which play spaces changed on the ground. Such an analysis contributes to our understanding of a critical period in the history of the children's playground and points to its complex place in wider social and political processes in this period. Unlike social housing, new towns and state-run industries, children's playgrounds remained publicly owned and communally funded despite a wider shift from social democracy to market liberalism. At the same time, commercial involvement in shaping the form and function of the playground continued and a wider range of play equipment manufacturers promoted their products and services. National government disinterest in playground provision was briefly punctuated by short-lived policy attention and dedicated funding in the 1970s and 2000s, while the fortunes of the playground would remain closely tied to the status, ambition and finances of local government throughout the period.

The power of play

The 1970s saw a renewed and widespread general interest in both childhood and play. Progressive educationalists, rooted in interwar ideas about psychology and child development, reached a much wider audience. For example, A.S. Neill's hugely influential book *Summerhill* achieved both considerable sales and widespread publicity in Britain and internationally for its promotion of childhood freedom and the role of play in education.[3] As we shall see later in this section, central government departments commissioned research into children's play and issued guidance on play space provision in a belief that play could help to achieve wider policy ambitions. At the same time, psychology was joined by sociology in trying to make sense of human nature and the place of children in society and the environment.[4] By the 1970s, play could seemingly provide evidence to explain a wide range of biological, behavioural and social phenomena, from its evolutionary role in animals and humans, through physical and social development, to its significance in the progress of western civilisation.[5]

In addition to its analytic potential, play was increasingly seen as not just a healthy but also a medically therapeutic activity. The gardens and grounds of asylums and other medical institutions had long performed a therapeutic function, providing space for open air convalescence and interaction with nature and horticulture. From the 1970s, childhood play was nurtured in such spaces too. Hospitals began to encourage inpatient children to play, nurses were trained to support playful activities and the Department of Health issued a circular to encourage, although not fund, play in hospitals.[6] In 1972, an outdoor play space that included climbing structures, a pond and a grazing goat was designed and created for Stoke Lyne Hospital by students from Exeter College of Art.[7] At Farleigh Hospital, near Bristol, an adventure playground was created for its psychiatric patients, although it could hardly compensate for brutal failings in care at the institution.[8] In 1970, Marjory Allen was involved in setting up an adventure playground in Chelsea, where disabled children and their siblings and friends could play together, followed by a wider association to support similar sites elsewhere.[9]

Despite the spread of these semi-public facilities for children's play in various medicalised environments, the lack of public play space was still seen as a problem, particularly in relation to new forms of housing. A study into family life on housing estates in Leeds, London and Liverpool found that the problem of high-rise living, combined with inadequate playground provision 'may well amount to a process likely to impair the normal personality development of the children affected'.[10] Play and the playground continued to be seen, for the time being at least, as important vectors for healthy child development. Such trends were evident in planning policy, where town planners continued to promote the principle of the playground as a device of childhood wellbeing. In 1961, the Parker Morris report, most well known for establishing internal space standards for council housing, also made recommendations relating to the provision of play spaces. A 1968 double edition of the journal *Town and Country Planning* showed how planners could improve children's lives at home, at school and at play, and emphasised the potential benefits of children's participation in planning for the future.[11]

The policy aspirations for planned play provision were implemented most notably and comprehensively in the new towns, where local authorities were replaced by semi-autonomous development corporations charged with making the purpose-built settlements a reality. Planners working on the development of Milton Keynes in 1973, adopted a particularly optimistic tone. Toddlers' play spaces close to home, communal playgrounds in parks, adventure play centres and a children's play officer would provide playful leisure opportunities to meet the needs of children and their families.[12] In Basildon, car-free public spaces were dotted with sculptural play equipment (Figure 5.1), while the provision of play space within residential areas was a strategic objective for planners in Harlow too.[13] Even where housing schemes were less extensive and involved a smaller extension to an existing community, children's play could still be central to

Figure 5.1: Open space with children's play area, Basildon, by S. Lambert, 1967, Architectural Press Archive / RIBA Collections, RIBA63840.

their design and layout. New estates provided a significant improvement in housing conditions and often afforded more space and greater freedom for play. For example, photographs of the Middlefield Lane estate in Gainsborough, Lincolnshire, show how children regularly used communal areas, so that while they 'were not quite places *for* children, they were child-centred in the hope of fostering children's wellbeing'.[14] Play space provision remained important in established urban areas too. In Waltham Forest in east London, the council's 1977 Corporate Plan placed a high priority on creating additional playgrounds, with eighteen new play spaces planned for parks, housing estates and education sites.[15] The playground was still an integral part of visions for a better urban environment and the provision of dedicated space for play remained an important aspect of an optimistic approach to planning new communities during the 1970s.

While new estates may have been positive spaces for some children, there were also increasing critiques of planning orthodoxy and its emphasis on creating planned spaces for play.

By the 1970s, planning had not made the urban world anew as its early advocates had often hoped. Many people still lived in dilapidated housing, in neighbourhoods that had either not been rebuilt after the war or were in the middle of slow rebuilding programmes. Over 240 local civic societies came together to describe a resulting 'urban wasteland' in many parts of the country, including Surrey Docks in London, Glasgow's east end and St Radigund's in Canterbury.[16] The documentary photographer Nick Hedges made this strikingly clear in his work for the homelessness charity Shelter (1968–72) and in a subsequent exhibition commissioned by the Royal Town Planning Institute at the Institute of Contemporary Arts in London.[17] His evocative images highlighted the enduring resourcefulness and adaptability of children at play, even as their surroundings decayed. A child playing on a small, enclosed balcony at the top of a monumental tower block or children playing on broken swings among crumbling buildings were very different from planners' utopian hopes for redevelopment schemes and the playgrounds that accompanied them (Figure 5.2).

Hedges's creative response to the problems came on the back of growing criticism among academics and journalists about modern planning

Figure 5.2: Swinging in a derelict playground, Newcastle by Nick Hedges, 1971, © nickhedgesphotography.co.uk.

and its consequences. Perhaps the most influential critique of planning and the places it was creating was *The Death and Life of Great American Cities* by the journalist Jane Jacobs, which offered a damning attack on the principles of modern city planning orthodoxy. Jacobs argued that planners had long been fixated by the ideas of Howard, Corbusier and others about how cities *ought* to work, rather than seeking to understand how they actually worked in practice through the everyday lives of ordinary people. Rather than the planners' aerial perspective, Jacobs favoured a view of the city from the sidewalk. She argued that the street did not represent the problematic space so often ascribed by planners, nor were dedicated spaces for play inherently any better: 'how nonsensical is the fantasy that playgrounds and parks are automatically OK places for children, and streets are automatically not OK places for children'. She mocked the 'grass fetishes' of park advocates and the 'science fiction nonsense' that green spaces somehow represented the lungs of the city. Instead, she argued that city streets had long possessed an important social function as sites of neighbourliness and community interaction. For children in particular, the street offered collective adult supervision, a variety of ways to play, space for imagination and creativity and opportunities to learn about adult society through imitation. She felt that children needed an 'unspecialized outdoor home base from which to play, to hang around in, and to help form their notions of the world' and the street was the best place for that to happen. In contrast, downgrading the street and removing children was the 'most mischievous and destructive idea in orthodox city planning'.[18]

Although initially published in the USA and drawing on her experience of living in New York, *Death and Life of Great American Cities* became hugely influential. It inspired others to question long-held assumptions about planning and contributed to a degree of introspection within the planning profession in Britain. Writing in *New Society* in 1969, a group of British academics, architects and critics considered what would happen if there was no planning at all, calling instead for experiments in 'non-planning'.[19] In addition, as wider political and public consensus about the authority of the planner dissolved in the 1970s and 1980s, there was considerable debate within the planning profession about its future. Nathaniel Lichfield, noted academic and planning consultant, felt that the system needed to be overhauled to ensure it met the needs of children.[20] Others from within the profession questioned planners' ability to mediate between society and the environment, casting doubt on the adequacy of the system, its philosophical foundations and relationship with society at large, leading to considerable defensiveness and resistance to change.[21] The conviction among planners, architects and politicians that modern approaches to the reconstruction of the city, including the provision of playgrounds, heralded a bright new future for society was increasingly being renounced by the same people that had endorsed it just a decade earlier.[22]

This existential challenge to town planning was exacerbated by an increasing awareness and sensitivity to the lived experience of city dwellers, as the social sciences became increasingly influential in both academia and more widely. Like Jacobs's work, Kevin Lynch's influential book *The Image of the City* had sought to shift approaches to the city from the bird's-eye view of the planner to that of the person on the street.[23] Psychologists, sociologists, geographers and others were subsequently inspired to study the everyday lived experience of the city and the impact of the environment on behaviour, including among children.[24] The concept of territorial 'home range', borrowed from animal ecology, was of particular interest as scholars sought to understand the ways in which children made use of the urban environment.[25] Most significantly, sociologists were increasingly attempting to understand the changing place of children in the city. Play, the playground and its relation to new forms of housing proved to be an important testing ground for new sociological research methods.

This focus on urban social change had its roots in earlier interest in day-to-day lived experience and attempts at more participatory forms of urban planning. As early as 1936 Elizabeth Denby, the housing consultant and friend of Marjory Allen, had demonstrated an interest in the views of residents in new housing schemes, even if her subsequent designs did not necessarily live up to future occupants' expectations.[26] The creation and development of Mass Observation in the late 1930s helped to foster a greater awareness of the everyday experiences of working-class city dwellers in particular. After the war, the Building Research Station continued to study user satisfaction with new forms of social housing, while sociologists investigated the social consequences of rehousing schemes, particularly the impact on residents' sense of community.[27] However, children's play only occasionally featured in these initial assessments. A 1954 study of the Lansbury estate found children largely playing in the street, while the NPFA's *Playgrounds for Blocks of Flats* (1953) assessed the quality of new play provision mainly by questioning those responsible for creating it.

One of the earliest attempts to adopt new sociological methods and position children more centrally was by Margaret Willis, a pioneering researcher employed from the early 1950s by the LCC architects' department. In *High Blocks of Flats*, a study of nine council estates, Willis surveyed families to understand their experience of living in high-rise homes, including the impact on children's play. She found that younger children living on higher floors were often kept inside rather than being allowed out to play and that where play spaces were provided they were often inadequate. Willis concluded that families with younger children should be housed on the lower floors of high blocks, so that it was easier for younger residents to play outside, a call that would be repeated in subsequent reports in other cities.[28] In a follow-up study of young children's play on four estates, Willis found that few children used the playground frequently, with many preferring to play on the service roads, grass areas and in the entrances to buildings. When children did play in the playground, the sandpit was by far the most popular amenity among children. In a specific study of seven sandpits, Willis concluded that such amenities were an important playground feature on high-density estates, but that many parents expressed anxiety about the unhygienic and unhealthy nature of sand.[29] In the two decades after Willis's pioneering work, which remained unpublished due to departmental hierarchies and bureaucratic protocols, there were regular sociological studies of the relationship between children and the urban environment; three in particular stand out for their focus on children and play provision.[30]

The first, *Two to Five in High Flats*, was published in 1961. Written and researched by the sociologist Joan Maizels (and supervised by a committee that included Marjory Allen and Margaret Thatcher MP), it found that plenty of advice existed about children living in flats but that 'official practice had lamentably failed to keep pace with precept'.[31] In addition to its account of playgrounds for high blocks of flats, it examined the thoughts and experiences of 200 resident families and playground users, promoting the notion that children should have a greater influence in the places they were expected to play, even if this was mediated through their parents. Its findings also demonstrate how wider debates about childhood, play and public space were being worked through by individual families. It showed how new flats in high-rise blocks created better living conditions, but also disrupted earlier patterns of play that had centred on the street outside the home. With most of the families visiting parks and playgrounds only occasionally, the physical, visual and psychological disconnect between a high flat and ground level estate playground was problematic for both parents and campaigners.

Five years later, the Building Research Station published a second notable study, *Children's Play on Housing Estates* (1966) by the sociologist Vere Hole. She utilised a range of techniques, including observation, timelapse cameras and film, to better understand how children played on new housing developments and what use they made of playgrounds and other landscape features. The study sought to uncover the play habits and preferences of children so that the adequacy of existing playgrounds could be assessed against their lived experience and needs, using scientific techniques.[32] If children's play could be properly understood, then perhaps designated play provision could be adapted to engage more children for more of the time.

Hole's findings showed that children played in unusual ways and in different places to those previously imagined by play space campaigners. Of the 5,494 children observed, most spent their time taking part in sedentary but highly sociable play, including sitting, standing, watching and talking, leading Hole to suggest that the playground functioned as a site for children's behaviour patterns that were not dissimilar to those of adults and the local pub. The playground and equipment often acted as a focal point for social gathering, where children would join friends or seek companions, and fifty per cent of children had soon left the playground to play elsewhere on the estate. Left to their own devices, many children sought out sociable encounters rather than physical excitement or exercise, with little observable difference between girls and boys. Hole eloquently described a picture of play 'which is restless, changing, where groups coalesce and dissolve but where there is an underlying element of more continuous activity or repose'.[33]

The research also highlighted the differences between children's preferences and adult expectations. While in the playground, the sandpit, paddling pool and swings were the facilities of choice for children, while sculpture and architectural features afforded more pleasure to adults than to younger play space users. Despite their children's demonstrated preferences, most parents' criticism of estate play space focused on the lack of orthodox, manufactured playground equipment and Hole found that they displayed little awareness of contemporary theories about play or play space provision. Most tellingly, Hole found that play space was just about holding out against increasing demands for car parking and that the start of children's television at 5 p.m. saw most children disappear from public spaces altogether to watch programmes such as *Blue Peter*, *Jackanory* and *The Magic Roundabout*.

A third key study into the ways children responded to the urban environment was carried out by Anthea Holme and Peter Massie and published as *Children's Play* in 1970.[34] Whereas Maizels and Hole had concentrated on new housing estates, Holme and Massie focused specifically on playgrounds, motivated by the sense that not enough was known about children and their play. To remedy this, they sought to provide documented evidence for planners and play providers, local authorities and designers. They contrasted play provision in an old neighbourhood in

Southwark and in the new town of Stevenage, surveyed 467 playgrounds across 19 local authority areas, interviewed parents and recorded the play activities of 1,800 children. Their research provides a useful snapshot of play provision in 1970 and the numerical significance of different typologies. In the nineteen study areas, fifty-four per cent of playgrounds were on housing estates, thirty-eight per cent were in parks, with eight per cent in other locations. Over seventy per cent of park playgrounds provided traditional manufactured equipment, while play spaces on housing estates were more likely to include a combination of traditional equipment and architectural, sculptural or improvised play features. Over eighty per cent of playgrounds had neither sand and water nor adult supervision.

Their research supported Marjory Allen's earlier complaint that the provision of play space was only rarely coordinated between departments within local authorities, with playgrounds variously the responsibility of parks staff, engineers, surveyors, housing officers, education officials, town clerks and development corporations. In Stevenage, responsibility was spread across five different departments, while in Swansea all playgrounds were the responsibility of just the parks department. Even when provision was coordinated, the ongoing influence of conservative values in relation to public parks continued to shape opportunities for play. The study found that many parks remained ordered and formal spaces where children were forbidden from walking on the grass, climbing trees or riding bikes. In addition, while most of the playgrounds studied were poorly designed and lacked stimuli for play, they were generally well maintained and clean. Alarmingly, where playgrounds were not well cared for, conditions were very bad. With no statutory responsibility to provide playgrounds, good quality provision that was well maintained relied upon the enthusiasm of individual officials, councillors and outside pressure groups, something the authors found to be haphazard at best.

Beyond these three studies, interest in the consequences of urban childhood grew. The NPFA's *Playgrounds for Blocks of Flats* had been reprinted six times by 1974. In addition, medical researchers were finding that flat dwellers suffered from a greater incidence of respiratory illnesses, while further sociological studies highlighted the difficulties facing urban children, including John and Elizabeth Newsom's longitudinal research in Nottingham, *Four Years Old in an Urban Community* (1968) and Pearl Jephcott's study of tall blocks in Glasgow, *Homes in High Flats* (1971).[35] Further research showed that high-rise living did not have an exclusive hold on poor-quality play provision and that low-rise council estates experienced problems too.[36] Evidence was building to support Holme and Massie's conclusion that a national policy was needed to ensure that high quality, special places for play were provided, with minimal restrictions and maximum play opportunities.

The publication of *Children at Play* in 1973 by the Department of the Environment perhaps seemed like the first step towards a national policy for play.[37] It raised the stakes in terms of the number of children studied, from the 200 families in *Two to Five in High Flats* and the 5,000 children observed for *Children's Play on Housing Estates*, to over 10,000 detailed observations of play in new and old housing areas. The report included a review of literature relating to children's play in the urban environment from the previous decade and interviews were conducted with children, parents and other adults in low-, medium- and high-rise housing in cities across the country. The study explored doorstep play, playgrounds, adventure playgrounds, wild areas, and children's 'unorthodox' play on garage roofs and elsewhere. It repeated earlier suggestions that families with children should be accommodated in houses or ground-floor flats with gardens, rather than on the upper floors of tall buildings. It also acknowledged that children did not solely play in playgrounds and so the wider housing environment needed to be able to withstand this playful use. It highlighted the work of the landscape architect Mary Mitchell in designing successful play spaces in Blackburn and provided images of well-planted playgrounds incorporating trees and shrubs. Mia Kellmer Pringle, psychologist and director of the National Children's Bureau, contributed as a consultant advisor to the study team, helping to ensure its child-focused approach to play. As a result, the report appears to be a comprehensive study of children's playful activity and an effective digest of the latest thinking on children and their play in the urban environment, in many ways a model of best practice.

Curiously, the tone of the document, its detailed suggestions and the images it used are completely at odds with its final design recommendations, which cover just one out of one hundred pages. Transposed word for word from Circular 79/72, a joint directive on play space issued by the Department of the Environment and Welsh Office a year earlier, the recommendations in *Children at Play* were based on a highly conservative understanding of the playground and its form. The circular stated that play spaces should be equipped with items from a shortlist of traditional manufactured equipment, including the swing, slide, climbing frame, see-saw, merry-go-round and rocking horse. In addition, it required surfacing to be hardwearing and existing trees to be retained only where possible. It applied specifically to local authority housing developments and unlike earlier communiqués it set a standard amount of play space, three square meters, and dedicated additional funding, £400 (£18) per child, to cover the cost of play space construction. The circular did not discuss imaginative, creative or adventurous play provision, the need for more trees, shrubs, flowers or other landscape features, nor the involvement of play leaders or specialist designers.[38]

How can this apparent mismatch between Circular 79/72 and the wider tone of *Children at Play* be reconciled, particularly given that both were a product of the same government department? On the one hand, the circular followed a long tradition of indifference towards play provision by central government, which had hardly mentioned, let alone endorsed, the creation of playgrounds over the previous two decades. Where play space was mentioned in government documents, it was generally in relation to housing policy. The Ministry of Health's 1944 *Housing Manual* did not mention play provision and the 1949 manual simply suggested playgrounds 'might' be provided.[39] In the 1950s, Conservative governments primarily sought to reduce the cost of housing provision through economy in the use of land, rather than improving the quality of estate amenities.[40] The 1957 *Housing Handbook* was highly dismissive of playgrounds, stating that there was a lack of research into the subject and that the approach to play provision advocated by campaigners such as the NPFA was unduly costly and therefore not strongly supported.[41] In the mid-1960s, play space was once again eligible for central government housing subsidy but as this information was hidden in an appendix to a circular on housing costs it hardly represented a ringing endorsement.[42] Instead, the government publicly stated that it would not insist on the provision of spaces for play.[43] An official account of the Ministry of Housing and its work, published in 1969, did not mention children nor play, despite asserting that a key role involved overseeing 'the urban environment and its impact on the citizen'.[44] For central government, the issue of play provision was a minor component of housing policy, something to be provided alongside clothes-drying areas and waste disposal, primarily at the discretion of local authorities.

On the other hand, by the late 1960s ministers were 'increasingly anxious to extend the provision of play spaces' in response to questions in Parliament and a wider appreciation of the importance of play.[45] In 1968, Ministry of Housing officials issued a guidance note to local authorities that included a short paper by the NPFA on imaginative playground design, including the noteworthy instruction: 'no old cars, play sculpture or other adult grotesqueries please'.[46] Even then, the covering note was explicitly clear that it solely represented the views and experiences of the authors and was in no way a government endorsement of the recommendations. Furthermore, the civil servants working to develop Circular 79/72 relied directly on a Wicksteed equipment catalogue to shape the instructions in the directive, rather than the NPFA note the department had previously shared with local authorities.[47] Ultimately, concerns about cost and administrative complexity dominated discussions between officials, rather than necessarily the needs of children when playing.[48] Mia Kellmer Pringle, consultant advisor on *Children at Play*, responded to an initial, confidential version of the circular by stating:

> I would not wish to be quoted as being in agreement with the provision outlined in your draft. Of course, it is a very reasonable first step and this may be all that can at present be afforded, but this is very different from saying that it is in any way adequate.[49]

Political pressure meant that civil servants had attempted to promote play provision, but the combined challenges of financial restraint and bureaucratic complexity limited the published standards to the bare minimum in the eyes of campaigners. Ministry officials acknowledged the likely opposition to the circular from the NPFA and Marjory Allen, but in the event the circular was far more widely criticised.[50]

While campaigners welcomed the dedicated funding that accompanied the directive, other aspects including its approach to play space provision were roundly condemned. The Inner London Education Authority felt that the low standards were totally inadequate and encouraged planners to do much more than the circular suggested in terms of space for play and its design.[51] For the deputy director of amenity services in Lambeth, the 'list of playspace equipment is sad, it might have been appropriate ten years ago but it isn't now'.[52] For one unnamed commentator, the circular lacked a definition of play space, the list of equipment was unimaginative, there was no mention of facilities such as water fountains or toilets and it had a narrow focus on equipment at the expense of other forms of play.[53] Alongside these criticisms, the NPFA were disgruntled not to have been consulted on the content of the circular and submitted a revised version that more closely resembled campaigners' thinking on play provision, but this was quickly dismissed by officials.[54]

Despite considerable sociological research and the ongoing efforts of campaigners to promote alternatives, the approach to play demonstrated by the circular was remarkably conservative, particularly given the findings from *Children at Play*, which were available to officials well before it was published. The problem partly stemmed from the differing expectations of a play space standard. For campaigners, a standard was meant to be aspirational, an ideal that providers should aim for in terms of the quantity and quality of play provision. Conversely, for central government officials the standards in Circular 79/72 were designed to be the bare minimum acceptable to attract subsidy, something that progressive local authorities would want to exceed. For critics, this disconnect in relation to the purpose of a standard meant that the government appeared to be significantly behind the times in terms of their approach to play.

In a review of research and guidance in 1976, Clare Cooper Marcus and Robin Moore concluded that while more was known about children's use of playgrounds, research findings had rarely been disseminated to those in central and local government, let alone shaped policy or implementation on the ground.[55] However, even researchers were selective in the findings they endorsed and extolled. Despite increased recognition that children played everywhere, there was still a sense that the playground was the place that children should play and the issue that needed to be solved related to the type of play spaces being provided.[56] In 1978, Moore asserted that 'the creation of childhood places cannot be left to chance or the vagaries of pressure groups; they must be deliberately fostered by planning, design, and management to satisfy basic human needs'.[57] In a similar vein, some park advocates in Britain continued to see play as a juvenile version of adult leisure and recreation, an activity that needed spaces and equipment for play in public parks at an appropriate frequency.[58] For others, this increased knowledge about the way that children played suggested that it was not the attractiveness or frequency of play provision that was the problem but rather, as Jane Jacobs had argued in the early 1960s, that the principle of the playground was unsound.

This sense that the playground concept was flawed developed further in two very different fields of thought. On the one hand, those who had long observed children at play recognised that play could happen everywhere and anywhere, that children adapted to whatever environment they happened to be in. For others, the conventional playground was a symbol of political oppression, a space that symbolically and often physically denied children the freedom of the city that belonged to them as much as adults. Turning to the former initially, the work of the folklorists, Iona and Peter Opie best represents this line of thought.

In their 1969 book *Children's Games in Street and Playground* they stated that

> during the past fifty years shelf-loads of books have been written instructing children in the games they ought to play, and some even instructing adults on how to instruct children in games they ought to play, but few attempts have been made to record the games children in fact play.[59]

Through observation and discussion with 10,000 children in England, Wales and Scotland, they collated details of children's spontaneous and self-directed outdoor games. They found that similar games were played across the country, but with regional variation in names and local tweaks to the rules. Games of chase that were called 'tig' in Scotland and the north of England, were 'tick' or 'tip' in north Wales and the west Midlands, 'touch' in south Wales, 'tag' around Bristol and 'he' in London and the south-east. The Opies felt that children's self-organised play demonstrated excitement, adventure, imagination and ways to opt out of the ordinary world. They concluded that 'where children are is where they play'. Significantly, this recognition that children played everywhere and anywhere fundamentally undermined the assumption that children required dedicated places to play. Children could adapt to their surroundings and enjoyed secret, wild places best, away from adult supervision, where juvenile community could thrive. As a result, the Opies were dismissive of both the traditional playground and its 'cage-like enclosures filled with junk by a local authority, the corners of recreation grounds stocked with swings and slides' and the adventure playground and its play leaders, 'the equivalent of creating Whipsnades for wild life'.[60] By focusing on children and their self-directed playful activities, the Opies found that the provision of playground spaces was something of a benign irrelevance to the social lives of children, just one of many spaces where children played and developed their own collective culture.

In contrast, for many left-wing radicals the playground was a highly visible feature of the wider exploitation and control of children by adults, an extension of the power exerted by men over women, the oppression of the working-class and attempts to enforce particular standards of behaviour in public space. With limited political power, children were seen to experience particular difficulties in their ensuing war with adults. For radicals, the conventional playground was a prime example of the way that adults had sought to control children, excluding them from the wider urban environment and limiting their behaviour through both designating space and the use of materials such as tarmac, ironmongery and fencing. There were calls of 'free the children, down with the playground' in response to efforts to enclose children's play.[61] In contrast, the adventure playground was often portrayed as an experimental space of childhood freedom and hope for a better society, a line of argument commonly associated with the anarchist writer Colin Ward.

In relation to Ward and others, the historian Mathew Thomson has argued that the late 1960s and early 1970s saw radical thinking and action in relation to landscapes for children. However, there is considerable evidence that points to the earlier development of radical ideas and action in relation to the children's playground in particular. As we saw in the previous chapter, the conceptions of childhood that were inherent in the adventure playground ideal were grounded in the beliefs of interwar progressive educationists and were most visibly introduced to a wider British public by Marjory Allen's *Picture Post* essay in 1946. In addition, Ward had been consistently promoting the adventure playground from the late 1950s. In 1958, he cited the adventure playground as a striking example of living anarchy, valuable both as a place in itself and as verification of libertarian rather than authoritarian values. He would make the case again in 1961, almost word for word, in a special edition of the journal *Anarchy* which focused on adventure playgrounds. The same text was largely reused for a chapter in *Anarchy in Action* (1973), this time reaching a wider audience and contributing to wider sociological investigations into urban childhood.[62] In suggesting earlier roots to radical ideas about children's play, this evidence does not diminish Thomson's suggestion that Ward's *Child in the City*, first published in 1978, represented a high point in radical thinking about urban childhood and the playground.

In *Child in the City* Ward built on the work of earlier sociological research and anarchist thinking to emphasise the extent to which children adapted the adult-imposed environment, where play provision operated on one plane and children on another. He would later write that *Child in the City* was intended as a celebration of children's resourcefulness.[63] To facilitate such ingenuity and imagination, he felt that city officials who were genuinely concerned for children should make the 'whole environment accessible to them, because whether invited or not, they are going to use the whole environment'.[64] In making a claim for the entire city for children, Ward differentiated his mission from that of other child advocates. He argued that 'if we seek a shared city, rather than a city where unwanted patches are set aside to contain children and their activities, our priorities are not quite the same as those of the crusaders for the child'.[65]

The artist and educator Simon Nicholson provides a good example of these alternative priorities and the development of a model for implementing them on the ground. Initially in the USA and later from the Open University in the UK, he sought to take children's play beyond the playground to create the shared city that Ward imagined.[66] His 'theory of loose parts' promoted greater child involvement in the design of both objects and places for play. In a phrase often quoted since, he stated that 'in any environment, both the degree of inventiveness and creativity, and the possibility of discovery, are directly proportional to the number and kind of variables in it'.[67] As such, meeting the needs of children required an adaptable and flexible urban environment at a variety of scales. At the smallest, Nicholson felt that individual children needed their day-to-day environment to include loose materials, such as water, fire, living objects and resources for building, seemingly inspired by the ethos of early adventure playgrounds. At a wider scale, the urban environment needed to be flexible enough to accommodate community involvement in shaping it, rather than being fixed by planners, architects and builders in turn. Widely cited by playworkers since, at the time his concept built on the ideas of anarchists and radicals and coincided with the practical and arduous efforts of local communities to reclaim space within the urban environment for play.

Campaigning and working for play

The difficulties that community activists would face in creating space for play embodied many aspects of the wider battle of ideas outlined in this chapter so far. For many parents, planners, health workers and campaigners, play retained its association with childhood wellbeing and dedicated play spaces were seen as a symbol of a healthy urban environment. At the same time, sociological research, radical thought and the existential challenge to planning unsettled not only traditional conceptions of the playground but also cast doubt on the need for dedicated play spaces at all. Despite mounting evidence that challenged orthodox play provision, central government intervened in the battle for ideas, overtly endorsing traditional conceptions of the playground. However, the playground scuffles were far from settled. During the 1970s and early 1980s they would be played out in struggling local authorities, radical play work, the expanding market for manufactured equipment and in sensationalist debates about safety. Before turning to these arenas, the next section explores the place of play in community politics and activism.

Community demand for play provision was not new in the 1970s, but it did achieve a higher profile and became embroiled in wider political struggles that moved beyond campaigns for a healthier urban environment to challenge established notions of democracy, inclusivity and civic responsibility. In 1936, ninety-two people had petitioned the local council for a playground in Okehampton, Devon, while a 1952 survey by the NPFA showed that parents wanted more places for their children to play.[68] By the 1960s, vocal demands for more and better play provision increased significantly and often constituted a key demand of community groups attempting to improve the urban environment in the face of apparent inaction by local authorities. Groups of protesting children also made for emotive coverage in local newspapers and signalled childhood agency in both political debate and the use of public space. In 1963, 200 children marched on Stockport town hall with a 2,000-signature petition, protesting at the lack of play space on their estate in Edgeley.[69] A few months later, a further 100 children marched from Brinnington on the other side of town, to protest at the lack of space for play in their neighbourhood.[70] In Lancaster, there were dramatic reports that hundreds of children had stopped the traffic and 'laid siege' to the council offices in another protest about play provision.[71] By 1970, housing officers in London felt under 'constantly increasing pressure, highlighted by petitions, threats of protest marches, representations from Tenants' Associations, MPs and our own members, social groups and the like, to provide bigger, better and more varied play facilities'.[72] Children's play had become a significant political issue at the local level, echoing the pressure felt in central government.

And just as photographers had recorded iconic images of postwar play on bombsites, film makers in the 1970s were drawn to attempts by children and their parents to secure space for play. A 1972 film, *It's Ours Whatever They Say*, documented the perseverance of residents living on the Lorraine estate in Islington, north London, to secure a safe playground for their children.[73] The newsreel style documentary included footage of children playing on a disused timber yard and the council's attempts to eject them, so that the site could be redeveloped for the more structured recreation associated with a scout hut and new housing. After threats of arrest, a protest march to the town hall, considerable local newspaper coverage, a renewed occupation of the site by children and finally the revelation of underhand behaviour by council officials, the film ends with residents securing the site as a space for their children to play. Another documentary film, *Do Something!*, included further coverage of the problems of children's play in Islington and highlighted its intersection with issues of poor housing, racial tension and a problematic relationship between the council and local residents.[74]

North London was not the only part of the capital where the playground became embroiled in direct action by local community groups. Residents in Notting Hill, west London, had experienced difficulties for many years, including riots in the late 1950s, exploitation and intimidation by private property owners and apparent indifference from the local authority. The social researcher Pearl Jephcott found that 'the local press reflects with dreary monotony the extent and variety of troubles which afflict the district'.[75] Serious gaps in the provision of play space for children were a consistent feature of studies into the area's problems and in time became a key demand of local activists, residents and protesting children.[76] As in Islington, the efforts of residents to exert more control of the spaces for play in their neighbourhood were evocatively captured on film. *The Battle for Powis Square*, filmed by Community Action Group in 1974 on portable video recorders, captured the ongoing attempts by the local community to open the square as a proper space for play and the dismissive and condescending attitude of local Conservative councillors.[77] The film documented residents' efforts to retain grass areas for play and to provide play workers, shelter, toys and other activities, at odds with the council's preference for an unadorned asphalt ground. Councillors imagined the playground as a tarmacked space, similar to surrounding streets but safe from traffic, while local residents envisaged a garden play space with supportive adults and appropriate facilities that would nurture the children of the neighbourhood.

These examples from Islington and Notting Hill highlight the extent to which the playground had become symbolic of a humane and liveable urban environment and attentive local government investment, not just among earlier advocates but also for parents and other local activists. The films' portrayal of angry parents and enlisted children demanding action from stony officials encapsulates at an individual level many of the wider tensions in the evolving battle of ideas relating to urban play. Further research into the production, distribution, viewing and reception of these films could provide useful insights into the relationships between parental activism, local politics, technology and changing conceptions of urban childhood and the way that these issues were played out in the playground. Residents and community groups certainly saw dedicated spaces for play as one way to improve the quality of their neighbourhood and the lives of their children. But neither a tarmac ground as imagined by the Kensington councillors, nor playgrounds with orthodox equipment were the answer. Instead, local activists sought to provide something closer to a community-focused play centre which incorporated free play, aspects of the natural environment and adult advocates for children's play, an approach which became synonymous with the developing play work profession.

The 1970s saw the expansion of an increasingly organised and professionalised play work sector, championed by the NPFA and grounded in an approach to urban childhood and play space inspired by the postwar adventure playground movement and subsequent community activism. And while the political urgency and associated publicity subsided from the late 1970s, on which more later, the story is important in wider accounts of the playground because the anarchic political values continued well beyond the decade in the culture of play workers and continues to influence present-day advocates for children and their play in public space.

From the 1960s, the NPFA had renewed its efforts to promote adult involvement in children's play. It sought to enhance the status of the emerging play leadership profession by organising training courses in play leadership in conjunction with the Institute of Park Administration and with input from many individuals involved in the early postwar experiments in adventure playgrounds.[78] The influential director of the NPFA's children and youth department, Drummond Abernethy (1913–88), had volunteered at Lollard Street adventure playground in the 1950s before spending two decades working for the NPFA, making it a nationally recognised centre for advocacy and advice in relation to progressive notions of play and play leadership.[79] After much debate over the name and purpose of the organisation, the Institute of Playleadership was established in 1970 by the NPFA, Institute of Park Administration, Marjory Allen and a number of early play workers.[80] The inauguration of the Institute was a notable step in the NPFA's shift away from orthodox visions of the playground towards more progressive, liberal notions of play. This reached a logical conclusion in the late 1970s when adverts for manufactured playground equipment were dropped from its journal, a process explored in more detail later in this section. But in associating itself with the values of more radical advocates for children's play, the NPFA's authority and reputation would be challenged by a conservative backlash against wider efforts to promote political and social liberation.

Even without this negative reaction, the Playleadership cause remained beset by fundamental uncertainty about the precise role of adults in shaping children's play and play environments. The increasingly organised and coordinated nature of the play work profession did not help to alleviate that doubt. In 1973, one Institute member felt the adventure playground and the role of play workers was ambiguous at best: 'for the team of six playleaders, most of them inexperienced, the problem was less clear-cut. Was the centre to be educational, recreational, therapeutic or a mixture of these? Each of us part teacher, doctor, counsellor, community worker, builder, cleaner, handyman.'[81] In addition, others felt that the overtly political stance of some workers was problematic.[82] The play worker was unenviably struggling to be everything to everyone, treading a fine line between community worker and political activist, something that more recent advocates have sought to resolve while also attempting to provide the profession with greater theoretical and practical foundations.[83]

Despite this ambiguity, the number of adventure playgrounds increased from the handful of postwar, short-term, experimental spaces.[84] A 1956 NPFA conference had brought together representatives from eight adventure playgrounds in Bristol, Cambridge, Grimsby, Hull, Liverpool and London, while Crawley sent apologies.[85] By 1962, the capital's four sites had formed the London Adventure Playground Association to coordinate their work. And by 1974 there were twenty full-time adventure playgrounds in London, three in Liverpool and four in Bristol.[86] Those with practical experience of the initial experiments shared their knowledge and sought to increase the status of the profession. There had been earlier descriptions of experiments in adventure play, such as John Barron Mays's account of Rathbone Street in Liverpool, but the 1970s saw a far greater number of publications as many play workers promoted their work and vocation.[87] In the space of a few years, Jack Lambert described his experiences of working at Parkhill adventure playground, Arvid Bengtsson provided a visual account of similar spaces around the world and Bernard McGovern penned advice on play leadership.[88]

Joe Benjamin's call for industry-sponsored adventure playgrounds might not have been realised, but in *In Search of Adventure* (1966) and *Grounds for Play* (1974) he cemented many of the enduring myths and tropes of subsequent play advocates. He repeated Allen's assertion that Sørensen invented the adventure playground and reiterated the problems of orthodox equipment. He promoted the apparent freedom of adventurous play, but at the same time failed to acknowledge the highly gendered assumptions about children's activities in such spaces.[89] Historian Krista Cowman has shown how the figure of the heroic male playleader, a character evident in many of the accounts mentioned above, significantly limited the potential of the adventure playground to challenge traditional gender norms, despite wider social changes that were altering the position of women in society.[90] In addition, not only did adventure playgrounds embody conservative gender values, but they were also becoming less radical in their approach to play.

In *Grounds for Play*, Benjamin lamented the shifting ethos of adventure playgrounds, as the continual processes of construction and destruction by children were gradually replaced by the labour of play workers with permanence in mind. Adult involvement in building improvised play structures was not new in the 1970s; the NPFA had produced and distributed plans for improvised play equipment in 1956, but the idea certainly gained momentum among play workers. By 1970, over 4,000 copies of the NPFA plan had been distributed and they continued to promote the idea in their journal.[91] Several books on do-it-yourself playgrounds were available, particularly from the USA, although the NPFA advised caution in their wider application in light of inadequate materials and fixings.[92] Despite Benjamin's observation that the gap between conventional and adventure playground provision was narrowing, radical play workers nonetheless saw themselves at odds with the providers of more orthodox play provision. Hughes would later recall that play workers' 'natural enemy' was the parks department and their 'lazy, adult-oriented and wasteful' approach to play provision.[93]

In some ways this was an unreasonable generalisation. The campaigning and advocacy work of the 1950s and 1960s had influenced the wider approach to play provision, in some local authorities at least. The London County Council, and from 1965 its replacement the Greater London Council (GLC), was lauded for its progressive approach to play. Its parks staff attempted to create bespoke, adventurous play spaces in attractive landscapes, with facilities for adults and children that would be flexible in their use, with layouts that could be properly maintained once opened.[94] In Battersea Park, a new playground included some traditional equipment such as swings and roundabouts but also provided wooden stockades, a broad slide with sandpit at the bottom and a miniature theatre, while a sand valley, mature trees and landscaping helped to create a 'natural' setting for the playground.[95] Its Play Parks scheme, running since 1959, was also praised.[96] It provided staff and additional creative play opportunities adjacent to more conventional spaces in fourteen public parks by the mid-1960s. Such spaces invariably included areas for den building, sandpits, a quiet area for imaginative play and equipment that included building blocks, water play, garden tools and toys. Marjory Allen would promote this approach, along with examples from Sweden, in her pamphlet *Play Parks* in 1964.[97]

Some landscape architects also continued to design imaginative and flexible spaces for play. Mary Mitchell won an award from the Civic Trust in recognition of her work on Birley Street Playground in Blackburn and her designs reached international audiences via the French modernist architecture magazine *L'Architecture d'Aujourd'hui*.[98] However, *Landscape Design*, the journal of the British landscape profession, continued to have few contributions relating to children's play during the 1970s.[99] This might be attributed to a greater awareness among designers of the need to think beyond the playground to create total environments for play, typified by Michael Brown's approach to housing landscapes. It could equally be the consequence of a plethora of guidance for landscape architects that already dealt with the issue, both in Europe and the USA. For Anne Beer, environmental planner at the University of Sheffield, the landscape profession had focused too much on meeting standards, like those set out in the government's *Children at Play* or the checklists promoted by some designers, rather than designing public spaces that responded to their surroundings and would meet the needs of children.[100]

London's Play Parks scheme hints at the problems associated with changing existing play spaces in response to the new ways of thinking that evolved in the 1960s and 1970s. The solidity and inertia of metal playground equipment not only limited opportunities for creative and flexible play in the minds of campaigners, but also made it difficult for designers and playground managers to adapt existing play spaces. Play Parks deflected this problem by providing additional play opportunities beyond the playground railings, whereas attempts to physically change individual play spaces would invariably take many years to realise. For example, after critical comments in the press and questions in Parliament in 1963, a survey of playground facilities by royal parks administrators concluded that 'we are lagging a good way behind the LCC in our children's play areas which at present are primarily designed for passive entertainment and do little to encourage spontaneous and creative play, which it is generally agreed is what should be aimed at.'[101] Rather than incrementally replace individual items of equipment, they decided to renew the Gloucester Green playground in the north-east corner of Regent's Park, which had originally been installed in the 1930s. A sum of £280,000 (£7,000) was nominally set aside and architects were instructed to design a scheme to replace the forty-year-old play space. After two overly expensive designs were rejected, a third, more affordable scheme received approval in 1969, work started on site in January 1971 and was completed that summer.[102] Not only had the project taken nine years from inception to completion but it was also hardly a demonstration of innovative playground design. The layout included large areas of hard surfacing and the retention of original equipment and iron railings, with concrete pipes and a fallen tree trunk the most obvious nod to current thinking.[103] Renewed again at a cost of over £1m in 2020, the present-day Gloucester Green playground demonstrates the slow pace at which playground thought has influenced spaces on the ground, while the incorporation of significant areas of naturalistic planting and the presence of donation pay terminals both speak to contemporary concerns about the nature of the urban environment and how we should pay for children's access to it.[104]

Away from the prestigious royal parks, playgrounds on municipal housing estates and in public green spaces faced a number of challenges during the 1970s. There was a shift in power and influence away from urban municipal authorities towards national government, as finances and policy making were increasingly centralised. As we have already seen, the centralisation of policy making meant that traditional ideas about playground provision were given significant weight by national government circulars and guidance. At the same time, local government reorganisation in the early 1970s increasingly meant that park and playground provision became the responsibility of more generalised recreation and amenity departments, much to the consternation of many in the parks profession.[105] As a result, the staff responsible for play provision often had little specialist experience or training, coming from backgrounds as diverse as libraries, sports or engineering.[106]

While reorganisation changed clerical structures it did little to change the day-to-day working of local authorities. For campaigners, the provision of play space remained beset by a lack of coordination. Municipal play space provision could still involve a range of officials from departments including the new recreation and amenity sections, as well as health, education, architecture and housing, while small rural parish councils often only employed a solitary clerk to manage all of their affairs, including playground provision. In central government nine different departments had some involvement in children's play provision in 1975.[107] The number of requests for a government guide to interdepartmental responsibility for play highlighted the ongoing uncertainty among local authority and voluntary organisations about where responsibility and direction lay. It also demonstrated the extent to which government did not seek to address this problem, with one Department of the Environment official keen to side-step 'the role of co-ordinator, which we have so far managed to avoid'.[108]

Beyond government policy and organisation, wider patterns and spaces of leisure were also changing, with free, communal provision including parks and playgrounds often sidelined by new spaces for leisure. The 1970s saw the dramatic growth of sports and leisure centres; historian Otto Saumarez Smith has argued that these often short-lived buildings represented both a continuation of municipal provision of facilities for health, but also combined the values of public health and commercial entertainment.[109] Some playgrounds responded by attempting to emulate this combination. At Wicksteed Park, by this point managed separately from the manufacturing company, the free playground and gardens were joined by paid-for attractions including crazy golf, dodgems, a rollercoaster, donkey rides, motorboat trips and a big wheel, all promoted in glossy colour brochures and other marketing materials.[110] Nature had not been completely relegated, although the small black-and-white pamphlet that highlighted its existence in the park was underwhelming in comparison to other publicity materials.[111] Nor was Wicksteed Park immune from the organisational troubles facing local authorities, with management consultants called in to restructure the operation of the site, complete with a new organogram structure, improved budgetary procedures and portion control in the cafeteria (the latter presumably for financial rather than health reasons).[112]

Play equipment manufacturers also attempted to adapt to changing attitudes to play and responded to some of the criticism levelled by more radical advocates. Companies sought to demonstrate how their products could fit into new conceptions of the playground, while a few railed publicly against criticism of their products. Unsurprisingly, all continued to promote the principle of the playground. Wicksteed & Co. produced landscape models to display their playground design expertise and ability to create undulating play landscapes that incorporated planting and sculptural features.[113] A concerted sales drive saw them diversify their advertising to include *Caravan, Chalet and Camp Site Operator, Council Equipment and Building News* and *Education Equipment*, as well as advertising supplements in regional newspapers.[114] They trialled 'Swedish-inspired' climbing structures and spacecraft roundabouts in an effort to demonstrate their ability to provide for imaginative play.[115] Their equipment catalogue emphasised the health and happiness that their products could deliver.[116]

Children were also a less regular feature of the text and images used to create manufacturers' advertisements and were sometimes missing altogether. Hunt & Son and Wicksteed & Co. implicitly acknowledged that children were not in reality their customers, but rather the municipal officials responsible for installing and maintaining playgrounds. As a result, their adverts emphasised the dependability and longevity of products, comprehensive aftersales service and included images of factories and maintenance vans rather than children playing.[117] Hunt & Son even took the unusual step of paying for advertising space to issue an open letter that responded to the comments of an unnamed critic. The company protested that government-imposed purchase tax stultified invention in equipment design, argued that popularity among children trumped adults' aesthetic considerations and that if their products were not popular or necessary that they would be out of business.[118]

In fact, the opposite seemed to be the case as the number of companies competing to sell playground equipment increased significantly from the late 1960s.[119] The three well-established manufacturers, Wicksteed, Hirst and Hunt, and their traditional equipment faced increasing competition as a wide range of new suppliers promoted alternative playground products that incorporated new materials and technologies. Bowen Associates' 'triodetic playdome' was an early example, a domed climbing frame made from aluminium tubes.[120] Playstyle introduced playcubes, a 'modular play system' comprising fourteen-sided, interconnected plastic polyhedrons and designed in conjunction with 'leading educationalists, child psychologists, playgroup leaders, playground designers ... and of course children'.[121] Recticel-Sutcliffe patented a new safety seat for swings, made of foam and rubber, to replace traditional wooden types (and even appeared on the TV series *Tomorrow's World* as part of a feature on how children might play in the future).[122] SMP Landscapes won a Design Council Award for their products, the first playground equipment manufacturer to do so, and their 'intensive use' play space in Leyton, east London, included a helicopter-shaped climbing frame and log cabin slide.[123] Other companies attempting to take a share of the playground equipment market included Record, Furnitubes, Massey & Harris, Tyneside Engineering, GLT Products, Kidstuff, Rentaplay, Sportsmark and Gilbert & Gilbert.[124] By 1979, the trustees of Wicksteed Park had asked SMP to take on an £88,000 (£11,000) project to redesign the original playground, an embarrassing indication of the extent to which Wicksteed & Co. and their products were seen to be increasingly antiquated and out of touch.[125]

The trade journals cashed in on the resulting demand for advertising space and in addition to the regular adverts from manufacturers they published long articles promoting equipment companies and their products. Previously, the publications had relied on contributions from parks administrators, landscape architects and sometimes playground campaigners to provide content for their pages. By the 1970s, these discursive or polemic articles had largely disappeared and were replaced by advertorial content, invariably written by a 'special correspondent' and exclusively based on information and images provided by the commercial equipment manufacturers.[126] In doing so, the companies collectively re-established themselves as the authority on public play provision for the parks profession and any debate about the function and form of the playground largely disappeared from trade journals like *Parks Administration* and *Parks and Recreation*.

For the NPFA and their journal *Playing Fields*, this was increasingly problematic. In the past, *Playing Fields* had included regular adverts from equipment manufacturers, alongside discussion about the provision and design of play spaces. By the 1970s, both the organisation and journal were resolutely advocating more progressive approaches to play provision, including adventure playgrounds, play parks, professional leadership and well-designed playgrounds, all of which constituted implicit criticism of more orthodox provision. In addition to the journal, the NPFA information centre was providing over forty publications on innovative play space design and leadership, including their own pamphlet series, information from the International Playground Association, as well as key texts by Allen, Bengtsson and others.[127] Partly to deal with this contradiction, the NPFA ceased publication of *Playing Fields* in 1976 and replaced it with *Play Times*, a more accessible magazine-style periodical that adopted an unequivocal approach to campaigning and no longer included advertising from equipment manufacturers.[128] But just as the NPFA wholeheartedly adopted the ideas and attitudes of the progressive playground movement, there was a corresponding increase in wider concern about child safety, which at times descended into panic.

Danger and decay

The debates about childhood freedom within and beyond the playground collided with other forces that increasingly characterised urban play provision as a problematic example of wider social and environmental malaise. In an article in the *Municipal Review*, Drummond Abernethy of the NPFA concluded that some parts of the country were providing good places to play, including in Bristol, Stevenage and Islington, but that elsewhere many play providers were failing in their responsibilities. In the same piece, the Bishop of Stepney, Trevor Huddleston, argued that many local authorities had got their priorities badly wrong, suggesting that 'it is not through malice or evil intent I am sure, but they have been overtaken by the motor-car and other matters so that play provision has been left behind'.[129] However, it was no longer the motor car that dominated the rhetoric of those concerned with children's safety as it had done for many playground campaigners from the 1930s to the 1950s. It remained a concern, but increasingly significant was the perceived risk to children from the unlikely combination of paedophiles, pets and even playgrounds themselves.

As we saw in earlier chapters, the threat to children from abusive adults had been recorded by park staff for much of the playground's existence, but the issue was often obscured by an unwillingness to talk openly about such incidents. By 1968, the sociologists Elizabeth and John Newson found that parents in Nottingham were increasingly protective of their children in response to fears about the dangers of sexual molestation.[130] Mathew Thomson has argued that this anxiety moved from a background concern to a major public issue from the mid-1970s. There was high-profile coverage of sexual crimes in the media and a public and political backlash against a short-lived sympathy, including among some adventure playground advocates, for a conception of the paedophile as a child lover rather than child molester.[131] Attempts at fostering rational public discourse by groups such as the Paedophile Information Exchange, including through conference attendance and book publishing, generated a little sympathy from the social work trade press and some in the gay liberation movement. In contrast, the frenzied response of tabloid newspapers and campaigns by socially conservative activists, such as Mary Whitehouse, meant this fragile sympathy was short-lived.[132] And so, just as greater freedom in play was being promoted by the findings of sociological research, anarchic politics, community activism and play advocates, there was an opposing anxiety about the risks of unacceptable adult behaviour that added to the perceived dangers of the urban environment.

A further danger to children at play was associated with pets rather than people. As historian Neil Pemberton has shown, there was concern in Burnley and beyond about the threat to children's health posed by dog faeces, in particular the problems associated with infection by toxocara canis (also known as dog roundworm).[133] This concern had developed from a number of scientific studies in Britain and the USA which pointed to the potential risks.[134] One in particular, which found that a quarter of soil samples from British public parks included the parasite, proved to be particularly influential.[135] Detailed descriptions of the parasite's 'hard, horny jaws which enable it to burrow through human tissue' and consequences that included loss of sight in children resulted in urgent calls for action.[136] Coinciding with rising concern about rabies, national newspapers portrayed a crazed canine menace that threatened children and their health in places where they were supposed to be safe.[137] This dramatic reporting, which continued for over fifteen years, contributed to the burgeoning anxiety about children playing in public space.

Adding to this angst was an awareness that in addition to hazardous people and pets, playgrounds could be dangerous places too. Up to the 1950s, there was a sense among those advocating for playgrounds that the safety of children at play was primarily the responsibility of mothers. *Playgrounds for Blocks of Flats* (1953) repeated a well-established assumption when it declared that playgrounds for young children had to be located within sight of mothers in their homes. Inadequate supervision by parents was even prone to scorn from officials. For one parks superintendent, the 'question of unaccompanied toddlers in a playground is quite serious and very difficult to overcome if irresponsible parents permit their small children to roam at will'.[138] Court cases relating to potential local authority negligence in playground provision were reported in the trade press, although they also reinforced the idea that parents had responsibility for children's welfare while using play spaces.[139] There was also a growing sense that playground safety needed to be considered by officials, partly in relation to the design of manufactured equipment but also the type of surfaces used, especially concrete.[140]

The 1950s saw the first trials of rubber safety surfaces, initially in the USA as an experimental collaboration between the Akron Board of Education and the Firestone Tire and Rubber Company, but also subsequently in Britain too. In Akron, the self-proclaimed rubber capital of the world, waste rubber was chopped into small pieces and stuck to the playground surface where it was found to lessen the problem of skinned knees and, unlike grass, dried quickly after rain.[141] In St Pancras in London, the noted architect Frederick Gibberd worked with the British Rubber Development Board to include rubber surfacing in the playgrounds of a redevelopment scheme because it was thought that it would help to minimise injuries and provide a harder-wearing surface than grass.[142] In both cases the organisations involved suggest that the initial development of safety surfaces in playgrounds was as much about the creation of new commercial products that dealt with industrial waste – and incorporated rather vague notions of safety in their marketing rhetoric – as it was about a direct response to evidence of specific dangers in the playground.

In the 1960s, playground safety remained a concern, although mainly for the professionals managing play spaces rather than the public more widely. For the LCC, it resulted in both anxiety and confusion among officials. In response to playground accidents, including several deaths, the council attempted to limit the use of apparently dangerous equipment. By the early 1960s there was considerable uncertainty among officials about which items of equipment could or could not be used, as no formal resolution had been reached. Parks and housing officials had made ad hoc decisions to initially stop installing slides, climbing nets, rocking horses and giant strides and subsequently all moveable, mechanical equipment. As a result, LCC architects produced bespoke designs for immobile playground features, including play walls, a wooden tent, tubular steel climbing frame and playhouse.[143] However, councillors were unwilling to make this official policy, so that by 1962 officials found that 'no specific list of barred equipment can be traced'.[144] In addition, playground safety was not always a dominant concern for housing officials. The attitudes of residents, Tenant Associations and councillors to playground facilities meant that experimental equipment could appear and disappear very quickly, as happened on the Aboyne estate in Tooting.[145]

Despite evidence that children were much better at judging their exposure to risk than their parents, by the 1970s the issue of playground safety received widespread and increasingly sensational publicity, while the design of playground equipment became a topic for discussion in national newspapers.[146] The *British Medical Journal* reported on a study by Cynthia Illingworth at Sheffield Children's Hospital which analysed injuries sustained by children while using playground equipment. While acknowledging that 'many accidents were due to normal childhood rashness', the article nonetheless highlighted the fractures, lacerations, concussion and other injuries associated with using swings, slides and climbing frames.[147] In 1972, the *Guardian* reported on the dangers of moving equipment, including the roundabout, rocking horse, ocean wave and swings, and the gruesome injuries and deaths they could cause.[148] A year later, *The Times* reported on the dangers for children at play, particularly the danger from hard wooden swing seats. It found that hospital records showed swings caused thirty to forty per cent of playground accidents; in Leicester and Manchester, the summer holidays saw ten children a day being admitted to hospital for head injuries; in the Netherlands, twenty children died each year in playground accidents. *The Times* described these figures as a major problem.[149]

These concerns were seen in the local press too. The safety of the playground in Wicksteed Park received considerable publicity in local newspapers, after reports of 700 accidents each year and action by the local authority Public Health committee.[150] In east London, the Wapping Parents' Action Group lobbied the local council and their MP after a child was injured by a rocking horse in a playground near Green Bank.[151] Parks staff and equipment manufacturers were accused of ignorance or indifference to the problem when they asserted that the number and severity of injuries were both insignificant. With no national data on the problem of playground accidents, it was hard to establish the scale of the problem and estimates varied wildly. In 1976, newspapers reported 20,000 playground accidents each year.[152] Two years later, the newly formed Fair Play for Children campaign estimated that '150,000 to 250,000 children are hurt or killed in playground accidents every year' on hard surfaces or play equipment that was often a 'death trap'.[153]

These dramatically increasing (and largely unsubstantiated) numbers were accompanied by well-publicised problems on the ground too, particularly with aging equipment that had not been adequately monitored. After several items of playground equipment collapsed, the GLC undertook a detailed inspection of its playground apparatus.[154] Rather than relying on visual inspection by park keepers as it had done for many years, technical officers conducted detailed scientific assessments and tests. As a result, half of its playground equipment was condemned, including seventy per cent of its slides, affected by corroding materials and inadequate safety precautions. The assessment of corrosion, particularly decay that was internal or invisible, undoubtedly required expertise analysis and subsequent remediation. But in using technical staff to assess questions of safety, the GLC unintentionally instigated a binary approach to playground risk, where equipment and spaces were either safe or unsafe, rather than recognising that assessing risk involved a value judgement, or even that the educational or developmental benefits of a playful activity could outweigh the risks. In addition, there was a sense that not only had older equipment not been properly checked and maintained but that the design of some items fell short. The British Standard, created in 1959 apparently to ensure the safety of equipment (although also partially to simplify and promote export business), was no longer adequate for ensuring the welfare of children at play.[155] The Design Council felt that the existing standard was so woefully out of date that it promoted its own list of reputable suppliers of safe playground products.[156]

By 1977, the wider pressure to address playground safety was being felt in central government. After a high-profile rocking horse accident in his constituency, the Labour MP and Secretary of State for the Environment Peter Shore asked his officials to expedite the revision of the earlier Circular 79/72 and *Children at Play*, and to encourage the British Standard Institute (BSI) to revisit the 1959 standard with an emphasis on safety. Obstinately, officials remained reluctant to intervene in the matter, suggesting in a handwritten note that 'this is a good example of an area in which we should not be intervening'.[157] Despite this, and with ongoing pressure from the Fair Play campaign and Consumer Association, civil servants did attempt to update the earlier circular and design advice.[158] After a number of difficulties including opposition from equipment manufacturers and problems coordinating the work of the BSI and Department, an interim letter was sent to local authorities in 1978 asking them to focus on the safety of both new and existing play spaces and to establish a more methodical approach to inspection and maintenance.[159] Although the Fair Play campaign was 'delighted' at the new advice, more experienced play advocates including the NPFA and local authority staff disagreed with many of its detailed suggestions, which often seemed naïve and disproportionate.[160] One respondent felt that the concerns expressed by the Fair Play campaign, which had obviously influenced the content of the interim letter, were extreme, lacked evidence and that 'injuries in children's playgrounds in most places seem to be quite remarkably light and may well be lower than almost any of the other common situations in which children find themselves'.[161] Despite these observations, a preoccupation with safety was further embedded in playground discourse when the British Safety Council (formed in 1957 and usually concerned with industrial accidents) successfully made the case that the 1974 Health and Safety at Work Act applied to play provision, much to the surprise of government officials.[162] This was to prove particularly problematic for adventure playgrounds.

Within the wider battle for ideas that characterised play space discourse in the 1970s, adventure playgrounds had already shifted considerably from their original emphasis on freedom, open access and child-centred play. In 1974, play worker Joe Benjamin acknowledged that such spaces had become significantly less accessible; they were invariably fenced and locked up unless a play worker was present, so that children were limited as to when they could use them. In Grimsby, the adventure playground even topped its fencing with barbed wire.[163] In addition, playful constructions that had previously been built by children on adventure playgrounds were increasingly replaced by large climbing structures, made from scrap materials, and instigated, designed and built by adult play workers.[164] The application of health and safety legislation to playgrounds meant that child-led activities were even more constrained.

In 1979, the Health and Safety Executive's Factory Inspectors issued a Prohibition Notice to the voluntary management committee of Northumberland Road adventure playground in Southampton, temporarily closing the site until significant dangers were addressed. The features that had previously symbolised freedom and creativity, including scrap materials, self-built structures, ladders and platforms, were now classified by the inspectors as unreasonably dangerous and a threat to the health and safety of people using the playground.[165] Before the management committee could appeal or take any remedial action, the local authority repossessed the site and cleared the playground.[166] Similar events took place elsewhere. After safety inspections in Manchester, the local authority instructed the city's adventure playground association to close all of its sites, forty-eight hours before the school holidays started.[167] In Suffolk, St Edmundsbury borough council dismantled Puddlebrook adventure playground on safety grounds.[168] By 1980, when the NPFA published *Towards a Safer Adventure Playground* many play spaces had already been closed or lost their local authority funding.[169] At the same time, financial problems were not limited to adventure playgrounds and by the late 1970s play spaces more generally were being affected by the consequences of reduced local authority budgets.

Play provision had long been subject to the vagaries of wider economic circumstances and the associated impact on local government finances and priorities.[170] In the 1960s, the documentary photographer Robert Blomfield captured the vulnerability of the playground to these wider conditions. The swings in Edinburgh's Harrison Park, with their seats removed and other equipment missing, appear desolate in the mist in his photo from 1960 (Figure 5.3). By the late 1970s, however, the playground increasingly became associated with the wider decline of parks and council housing estates. Landscape historians have characterised the 1970s as the start of a gradually intensifying period of decline in public parks, caused in part by reductions in municipal funding and the associated reduction in the quality and quantity of maintenance, but also a response to the low profile of parks within newly reorganised municipal structures and changing leisure patterns.[171] Park keepers were jettisoned

Figure 5.3: Harrison Park, Edinburgh, by Robert Blomfield, 1960, © Robert Blomfield Photography.

and local authorities attempted to manage the decline of the Victorian park model.[172] In large cities, such as Liverpool, these national trends were compounded by local economic decline and persistent social problems, which meant that parks and their amenities dropped even further down the local political agenda.[173]

There was a sense among some commentators that public parks and the playgrounds within them were in crisis, increasingly obsolete, badly managed, expensive and underused.[174] But if the term 'crisis' implies a sense of calamity or urgency, it seems more likely that public spaces were experiencing a long, slow decline in response to incremental changes in management and as inspection and maintenance regimes were neglected.[175] In an attempt to shame authorities into taking action to improve matters, *Play Times* instigated a 'brick of the month' award in 1977 which highlighted the poor-quality design and non-existent maintenance of play spaces across the country. Birmingham, once praised for its play space designs, was criticised for an accumulation of notices prohibiting play on its estates.[176] A decaying concrete train in Portsmouth and a neglected playground in Neath, with rusting, seat-less swing frames and the remains of a 'dead' concrete giraffe, were lamentable winners.[177] Knowsley borough council seemed to achieve the greatest fall from grace, with playgrounds left to decay so that they were 'the worst the editor has seen in a decade of looking at playgrounds all over the country'.[178]

Housing areas fared little better as the pragmatic modernism of the council estate playground was often incrementally chipped away in response to the problems of providing parking space for cars and increasing indifference among politicians and officials. The modernist Quarry Hill flats in Leeds provides a good example of this process. The estate was designed in the 1930s to have five playgrounds, but when construction worked stopped prematurely in 1940 only three had been laid out and equipped. Despite postwar agitation by tenants, the land set aside for the two additional play spaces remained 'deserts of glass-strewn asphalt ... destitute of all furnishings'. In the 1960s, twelve per cent of playground space on the estate was reallocated to car parking and most grass areas were enclosed with fencing and children's play there forbidden. Unlike the grass, the three playgrounds were now without fencing and almost all of the equipment had been removed, dismantled by exasperated officials in response to continual hard use and occasions of 'wanton vandalism'. By the early 1970s, the estate's buildings were characterised as obsolete, while the external environment was labelled 'intolerable', and the decision was made by local councillors to demolish the entire estate.[179]

Playground monsters

After the 1979 general election, a new Conservative government exacerbated the decline of both park and housing play spaces. The new administration rescinded Circular 79/72, removed the accompanying funding and distanced itself from play provision altogether, emphasising in Parliament that it was up to local authorities to decide on appropriate play space arrangements.[180] The somewhat perplexing sight of the new prime minister, Margaret Thatcher, opening an adventure playground in London's east end might seem paradoxical, given her government's attitude towards local government and the place of the adventure playground in left-wing, anarchic politics. However, she adroitly used the opportunity to expound her views about the dehumanising effects of state intervention and advanced the possibilities of charitable action, praising the voluntary management committee that had established and funded the play space.[181] Behind the scenes, officials pragmatically concurred, suspicious that local authorities had often taken advantage of the earlier playground subsidy provided by the circular but had not always used it to provide play spaces on the ground.[182]

At the same time, the urban redevelopment projects of the postwar decades, in which children represented hope for a better society, were recast as dystopian environments where children were a significant cause of the social and physical decline. Whereas playground advocates had long emphasised the deleterious impact of the city, by this time children were conversely seen to wield significant power over their city surroundings through vandalism, graffiti and other antisocial behaviour. Rather than addressing the root causes of such acts, a Home Office report laid the blame squarely on young people, suggesting that reducing child density on estates by dispersing families with children was a potential solution.[183] In 1985, Alice Coleman and the Design Disadvantagement Team at King's College London controversially cast judgement on modern housing landscapes and the children that lived in them. In considering playgrounds and the 'hordes of anonymous children' that they attracted, Coleman argued that dedicated play spaces were closely associated with a deterioration in the quality of lives of all estate residents.[184] Against a number of measures, including the existence of graffiti, litter, damage, urine and faeces, she argued that playground 'absence is better than their presence' and even associated the existence of play spaces on estates with a higher likelihood that children would end up in the care of social services.[185] For Coleman, the answer was to remove playgrounds altogether, dividing the space up into private gardens for ground-floor residents. Although rebutted by other researchers, the shift from seeing children as victims of the urban environment to blaming them for its problems was echoed in the popular press too.[186] Subsequent newspaper reports vilified 'tiny vandals' and described the terror of living on council estates with 'playground monsters', rhetoric that would continue well into the 1990s.[187]

Furthermore, with no government guidance on play provision, disenchantment with council estate landscapes and the denigration of problematic children, advice on playground standards was increasingly reduced to a technical matter. The revised British Standard, *B.S. 5696 Play Equipment Intended for Permanent Installation Outdoors*, was published in 1979 and received considerable publicity in trade journals.[188] While part pertained to the construction of manufactured equipment, the Standard also now covered site layout, surfacing, maintenance and inspection regimes and placed particular emphasis on safety.[189] In addition, by focusing on the provision of traditional equipped play spaces it further legitimised that particular vision of the children's playground at the expense of one which emphasised freedom, creativity or interaction with nature as imagined by play campaigners.

As we have already seen, the number of companies supplying playground equipment expanded significantly, as did suppliers for rubber safety surfaces. The two experiments with rubber surfacing in the 1950s evolved so that 'safety surfaces' became a common feature of playground provision from the 1970s and offered a new business opportunity. In 1977, *Play Times* tried to promote a measured approach to the use of increasingly expensive surfacing, suggesting that grass, sand and wood chip could all provide a suitable playground surface.[190] In busy areas, asphalt might even be suitable and at around £37 (£3.60) per square metre relatively affordable. At £385 (£37) per square metre, rubber safety surfacing was over ten times more expensive, and offered little to children in terms of play opportunities, but nonetheless became an integral feature of later play space provision as vague notions of safety trumped both play value and value for money.

The significance of close relationships between local authority officials and equipment suppliers in reinforcing this approach to play provision is not clear. The bribery and fraud associated with municipal housing contracts uncovered during the Poulson scandal had undoubtedly raised important questions about the ethics of public officials and the aptitude of local authorities, but there was no explicit suggestion that this behaviour extended to other sectors.[191] There were, however, close relationships between playground equipment manufacturers and local authority politicians and officers.[192] It is not clear for instance why such senior politicians and officials, including the lord mayor, chair of the parks committee and the director of parks, from a small Midlands city all needed to visit the factory of Wicksteed & Co. to develop their plans for a new playground, something normally dealt with by junior staff.[193] Appropriate or not, these close business relationships only strengthened the place of the equipped playground in the practices of local authorities and invariably made manufacturers the first port of call when money was made available to create or enhance dedicated spaces for children's play.

As urban historian Guy Ortolano has suggested, 'on either side of the 1970s, history remained untidy'.[194] Despite broader shifts in the political landscape during the 1980s, away from social democracy and towards market liberalism, the public playground remained free to use and communally funded. Unlike social housing and state-run industries, ownership of public play spaces was not subject to high-profile privatisation. At the same time, local authorities continued to buy playground equipment from commercial suppliers, while also being compelled to outsource playground maintenance to private companies through compulsory competitive tendering. Privately run indoor soft play centres and play zones within shopping centres provided alternative spaces for play, at least for those families who could afford to access them.[195] Since its inception in the nineteenth century, the form and function of the children's playground had been shaped by the interaction of a wide range of actors. In the 1980s, local authorities continued to own and manage public spaces for play, but limits on municipal power, funding and social remit left playgrounds somewhat adrift in the urban landscape.

This situation continued for much of the 1990s. Central government involvement in play space provision remained limited and the equipped playground remained the dominant conception of a public space for children's play. In 1996, researchers at the University of Reading found widespread antipathy in government, with no coordination or mandatory responsibility for play provision.[196] The British Standard was revised again in 1997 and remained focused on the safety requirements and layout of manufactured playground equipment, as did guidance booklets published by RoSPA on creating and inspecting play spaces.[197] With sponsorship of RoSPA's booklets provided by equipment manufacturers, including Wicksteed and SMP, it is perhaps unsurprising that swings and roundabouts continued to dominate both professional and public conceptions of the playground.

The election of a Labour administration in 1997 did not initially mark a radical change in the approach of central government, but there was a renewed interest in the form and social function of children's play spaces among researchers and children's advocates. In 1999, a special edition of the academic journal *Built Environment* focused entirely on the problems and potential of the playground.[198] Contributors pointed to ongoing uncertainty among providers about why they provided playgrounds and children's limited opportunities to participate in play space design, but also the ways in which such dedicated public spaces provided important meeting places for adults and children alike. The papers also highlighted ongoing points of contention between researchers who emphasised the importance of qualified professionals in designing the public realm and play work inspired advocates who promoted children's agency in the urban environment.

At the same time, play advocates continued to lobby for policies and funding to improve children's public play opportunities. In 2000, the NPFA and Children's Play Council published *Best Play*, a report that lamented the loss of suitable public spaces for play and the associated impact of 'play deprivation' on children's physical and emotional development.[199] Beyond the perceived health benefits for individuals, advocates invested children's play with social and political power and positioned as a remedy for poverty, deprivation and antisocial behaviour, much as it had been for well over a century. Furthermore, the report asserted that appropriately designed public play provision would help children to become economically useful and socially responsible adult citizens. In the early twentieth century, play advocates had imagined that gymnastic exercise or playful excitement would deliver these outcomes. In the twenty-first century, campaigners continued to draw on rhetoric first developed by the adventure playground movement, emphasising children's freedom of expression, imagination and involvement.[200]

For the next decade, campaigners inspired by and drawn from the play work profession promoted the importance of play for children's wellbeing, encouraged more rational approaches to playground risks, and lobbied for public investment in play.[201] In time, governments across the UK responded with programmes and sometimes funding to promote children's playful environments. The Welsh government created a pioneering Play Policy in 2002, while the Play Strategy for Scotland was published in 2013.[202] Both governments have also placed a play sufficiency duty on local authorities in their jurisdictions, requiring an assessment of existing provision and the creation of new spaces for play where necessary. In England, the 2008 Play Strategy was accompanied by £235m in funding and design guidance that sought to create more naturalistic play spaces, with less emphasis on manufactured equipment, as part of a wider drive by central government to improve educational outcomes and reduce child poverty.[203] This high-profile intervention was short-lived and in 2010 a new coalition administration withdrew both the guidance and funding.

Despite this, the children's playground remained a feature of social and political debate in the twenty-first century. The playground has continued to provide a focal point for high-profile conversations about the inequitable impacts of financial austerity, the segregation of public space by housing tenure and social class, and even the 'intrusion' of wild animals into spaces ostensibly set aside for children's play.[204] This ongoing public interest is unsurprising given the playground's longstanding physical presence in the urban environment, the malleability of the playground concept and its place in collective assumptions about children's place in the city. Despite controversy and debate, more than 26,000 children's playgrounds are still owned and managed by over 400 public organisations, representing all tiers of government across the UK. In an unusual quirk of welfare provision, the Ministry of Defence is the largest single provider of children's playgrounds in the UK, alongside large London boroughs, unitary authorities and small parish councils.[205] This administrative complexity and dispersed responsibility goes some way to explaining why playground provision has changed relatively little in the last century, despite the efforts of campaigners and radical changes in our understanding of children's lives.

In summary, from the late 1960s a battle of ideas in relation to the playground pitched radical, anarchic notions of childhood freedom against an increasingly widespread but often unsubstantiated preoccupation with safety. Play workers pitted themselves against parks departments, while the places in which children were supposed to play made national headlines for all the wrong reasons. The act of playing continued to be seen as a healthy component of childhood, but the playground suffered an acute loss of purpose. The nineteenth-century children's gymnasium and twentieth-century orthodox playground had both been inscribed with an unequivocal mission; to deliver childhood health through physical exercise, excitement and interaction with the natural environment, even if this was not always achieved in practice. In contrast, the postwar debate about the form and function of the playground had unsettled this mission. Planners, and briefly central government, felt that dedicated play spaces were still a worthwhile investment, but as the foundations of the planning profession were challenged, the comprehensive redevelopment schemes that had previously provided space for playground creation fell out of favour. An earlier sense that playgrounds could contribute to the future of society through the health of children was replaced by less grand aims, as public spaces more generally struggled to find a place in the new world of leisure and altered attitudes to local government. In any case, sociologists had shown that children did not really use playgrounds and as a result they were increasingly seen as something of an irrelevance. Despite this, playgrounds in parks and on housing estates endured, increasingly the preserve of equipment manufacturers, more about commerce than child development, health or happiness. At the same time, adventure playgrounds became less about providing a public space for play and more akin to a radical community centre; a DIY version of the orthodox playground, built by adults rather than children as originally intended.

From the 1970s, an increasingly widespread anxiety about safety meant that all playground typologies were reduced to spaces where risk needed to be managed, a problem to be solved rather than a space of possibility and potential. Interaction with curated forms of nature, once a key rationale in visions for the playground, disappeared almost entirely from the late-twentieth-century playground debates. The landscape architecture profession, previously advocates for nature in the playground, stepped away from play space design for several decades. Less of a priority in the face of organisational change and financial stringency, many existing playgrounds were seemingly abandoned by authorities. Paradoxically, this meant that a wilder version of nature began to reclaim at least some urban play spaces. Just as the principle and form of the playground waned, the NPFA also lost its reputation and function as a source of authority and expertise in relation to play space. At its inauguration in 1926 senior politicians had lined up to support the cause, but by the 1970s ministers and officials felt it was amateurish and doubted its technical advice.[206] With central government no longer seeking to guide or fund play provision, the British Standard, with its traditional conception of the playground and corresponding emphasis on safety, became the mainstream source of advice on playground provision in the later twentieth century.

But perhaps this did not matter. One of the primary goals of interwar playground advocates had been to reduce the number of children being killed on the streets and by the 1970s this had largely been achieved. But despite campaigners' expectations, it was not the playground that had protected children from the dangers of playing among motor vehicles. Instead, children were increasingly confined to the home, private garden or commercial play centre, while the car had replaced the child in public space. Ironically, by the 1980s critics blamed the playground for this loss of freedom, asserting that such spaces represented an unreasonable attempt to control children's behaviour, irrelevant to children's needs and described by children themselves as boring.[207]

The broader shift from social democracy to market liberalism undoubtedly affected the management, maintenance and perception of public play spaces during the 1980s and 1990s. But as a site that had long balanced public provision and commercial products, this case study of the children's playground highlights the complex ways in which broader historical processes were played out on the ground. Despite a very brief interlude in the early twenty-first century, when government guidance and funding held promise for advocates of children's inclusive access to the public realm, the playground story is still dominated by manufactured equipment. And while central government once more shows little interest in play space provision, many local authorities remain committed to the principle of the orthodox playground. Considerable sums are spent maintaining and improving existing play spaces, while local planning policies invariably require the provision of play space in new housing developments. However, as attitudes to childhood have changed, the public playground no longer acts as an extension of the home and a site for unsupervised play. Instead the twenty-first-century playground represents a rather confused response to children's place in the city.

Notes

1. Lawrence Black and Hugh Pemberton, 'The Benighted Decade? Reassessing the 1970s', in *Reassessing 1970s Britain*, ed. Lawrence Black, Hugh Pemberton and Pat Thane (Manchester: Manchester University Press, 2013), p. 14.

2. Mathew Thomson, *Lost Freedom: The Landscape of the Child and the British Post-War Settlement* (Oxford: Oxford University Press, 2013).

3. A.S. Neill, *Summerhill: A Radical Approach to Child-Rearing* (Harmondsworth: Penguin, 1968).

4. Mathew Thomson, *Psychological Subjects: Identity, Culture, and Health in Twentieth-Century Britain* (Oxford: Oxford University Press, 2006); Mike Savage, *Identities and Social Change in Britain since 1940* (Oxford: Oxford University Press, 2010).

5. For an overview of the place of play in wider thinking see, for example, J.S. Bruner, A. Jolly and K. Sylva, eds., *Play: Its Role in Development and Evolution* (Harmondsworth: Penguin, 1976).

6. Susan Harvey, 'Play in Hospital', *Mental Health*, 24.3 (1965), 121–3; Chiswick Polytechnic, 'Hospital Play Specialist Course', 1973, National Archives, MH 152/134; Joint Board of Clinical Nursing Studies, 'Children's Play Panel', 1977, National Archives, DY 1/77; Department of Health and Social Security, *Play for Children in Hospital*, Circular HC(76)5 (London: HMSO, 1976).

7. Jane Bentley and Laura Freeman, 'Play Space: The Design, Research and Development of a Play Area for Courtenay Special School, Stoke Lyne Hospital, Exmouth', 1972, Museum of English Rural Life, Landscape Institute, Pamphlet 2870 Box 1/08.

8. National Playing Fields Association, 'Adventure Playgrounds at Farleigh Hospital', 1971, National Archives, CB 1/63; 'Farleigh Hospital', *British Medical Journal*, 2 (1970), 58–9.

9. 'Adventure Playground for Handicapped Children', 1970, Wellcome Collection, Robina Addis Archives PP/ADD/K/2/2.

10. Ministry of Housing and Local Government, *Families Living at High Density: A Study of Estates in Leeds, Liverpool and London*, Design Bulletin, 21 (London: HMSO, 1970), p. 35.

11. 'Children and Planning', *Town and Country Planning*, 36.10–11 (1968), 430–512.

12. Milton Keynes Development Corporation, 'Play in a New City', *Playing Fields*, 34.1 (1973), 26–31.

13. Sam Lambert, 'Housing, Laindon, Basildon, Essex: Open Space with Children's Play Area', 1967, RIBA Collections, AP Box 752; Frederick Gibberd, *Harlow: The Story of a New Town* (Stevenage: Harlow Development Corporation, 1980).

14. Ian Waites, '"One Big Playground for Kids": A Contextual Appraisal of Some 1970s Photographs of Children Hanging out on a Post-Second-World-War British Council Estate', *Childhood in the Past*, 11.2 (2018), 114–28 (p. 126).

15. M. Hart, 'Dual Use Education Playgrounds', *Parks and Recreation*, 43.8 (1978), 40–41.

16. Timothy Cantell, *Urban Wasteland: A Report on Land Lying Dormant in Cities, Towns and Villages in Britain* (London: Civic Trust, 1977).

17. Alison Hall, 'The Shelter Photographs 1968–1972: Nick Hedges, the Representation of the Homeless Child and a Photographic Archive' (unpublished thesis, University of Birmingham, 2016); Nick Hedges, Larry Herman and Ron McCormick, 'Problems in the City' (Institute of Contemporary Arts, London, 1975).

18. Jane Jacobs, *The Death and Life of Great American Cities: The Failure of Town Planning* (Harmondsworth: Penguin, 1964), pp. 91–101.

19. Reyner Banham and others, 'Non-Plan: An Experiment in Freedom', *New Society*, 20 March 1969, pp. 435–43.

20. 'Planning Standards Criticised', *The Guardian*, 17 March 1966, p. 2.

21. Sylvia Law, 'Planning and the Future: A Commentary on the Debate', *The Town Planning Review*, 48.4 (1977), 365–72.

22. Otto Saumarez Smith, 'The Inner City Crisis and the End of Urban Modernism in 1970s Britain', *Twentieth Century British History*, 27 (2016), 578–98.

23. Kevin Lynch, *The Image of the City* (Cambridge, MA: MIT Press, 1960).

24. See, for example, the thirteen-volume series Human Behaviour and Environment: Advances in Theory and Research (New York: Plenum Press), edited from 1976 to 1994 by the social psychologist Irwin Altman, and the journal *Children's Environment Quarterly*, 1984 to 1995.

25. Howard F. Andrews, 'Home Range and Urban Knowledge of School-Age Children', *Environment and Behavior*, 5.1 (1973), 73–86; Amos Rapoport, 'The Home Range of the Child', *Ekistics*, 45.272 (1978), 378.

26. Elizabeth Denby, 'Rehousing from the Slum Dweller's Point of View', *Journal of the Royal Institute of British Architects*, 44 (1936), 61–80; Elizabeth Darling, 'What the Tenants Think of Kensal House: Experts' Assumptions versus Inhabitants' Realities in the Modern Home', *Journal of Architectural Education*, 53.3 (2000), 167–77.

27. Vere Hole, 'Social Effects of Planned Rehousing', *The Town Planning Review*, 30.2 (1959), 161–73; Michael Young, 'Kinship and Family in East London', *Man*, 54.210 (1954), 137–9.

28. Margaret Willis, 'High Blocks of Flats: A Social Survey', 1955, London Metropolitan Archives, GLC/HG/HHM/12/S026A Toddlers' Playgrounds; John P. Macey, 'Problems of Flat Life', *Official Architecture and Planning*, 22.1 (1959), 35–8.

29. Margaret Willis, 'Toddlers' Playgrounds: An Enquiry into the Reactions of Mothers to the Experimental Play Equipment for Pre-School Aged Children', 1959, London Metropolitan Archive, GLC/HG/HHM/12/S026A; Margaret Willis, 'Sandpits: A Social Survey', 1951, University of Edinburgh, PJM/LCC/D/6.

30. Ruth Lang, 'The Sociologist within: Margaret Willis and the London County Council Architect's Department' (presented at the Architecture and Bureaucracy: Entangled Sites of Knowledge Production and Exchange, Brussels, 2019).

31. Joan Maizels, *Two to Five in High Flats* (London: Housing Centre, 1961), p. 1.

32. Vere Hole, *Children's Play on Housing Estates*, National Building Studies Research Paper, 39 (London: HMSO, 1966).

33. Hole, *Children's Play on Housing Estates*, p. 23.

34. Anthea Holme and Peter Massie, *Children's Play: A Study of Needs and Opportunities* (London: Michael Joseph, 1970).

35. D.M. Fanning, 'Families in Flats', *British Medical Journal*, 4 (1967), 382–6; Pearl Jephcott, *Homes in High Flats* (Edinburgh: Oliver and Boyd, 1971); John Newson and Elizabeth Newson, *Four Years Old in an Urban Community* (London: Allen & Unwin, 1968).

36. A.T. Blowers, 'Council Housing: The Social Implications of Layout and Design in an Urban Fringe Estate', *The Town Planning Review*, 41.1 (1970), 80–92.

37. Department of the Environment, *Children at Play*, Design Bulletin, 27 (London: HMSO, 1973).

38. Department of the Environment and Welsh Office, *Children's Playspace*, Circular 79/72 and 165/72 (London: HMSO, 1972).

39. Ministry of Health, 'Housing Manual', 1944, National Archives, HLG 110/10; Ministry of Health, *Housing Manual* (London: HMSO, 1949).

40. Ministry of Housing and Local Government, *Houses 1952: Second Supplement to the Housing Manual 1949* (London: HMSO, 1952); Ministry of Housing and Local Government, *Houses 1953: Third Supplement to the Housing Manual 1949* (London: HMSO, 1953); Ministry of Housing and Local Government, *Houses 1953*, Circular 54/53 (London: HMSO, 1953).

41. Ministry of Housing and Local Government, 'Housing Handbook', 1957, National Archives, HLG 31/11.

42. Ministry of Housing and Local Government, *Housing Standards, Costs and Subsidies*, Circular 36/67 (London: HMSO, 1967).

43. *House of Commons Debate, 2 February 1965, Vol.705, Col.873, Housing Estates (Children's Play Spaces)* (Hansard, 1965).

44. Evelyn Sharp, *The Ministry of Housing and Local Government* (London: Allen & Unwin, 1969).

45. W.C. Ulrich to P.R.O.s, 'NPFA Advice on Children's Playgrounds', 1968, National Archives, AT 54/24.

46. W.D. Abernethy, 'Children's Playgrounds', 1968, p. 3, National Archives, AT 54/24.

47. 'Play Space Standards: List of Equipment Taken from Wicksteed Catalogue', 1971, National Archives, AT 54/24.

48. T.M. Heiser, 'Note for File: Play Space', 18 November 1971, National Archives, AT 54/24.

49. Mia Kellmer Pringle to J. Littlewood, 'Standards', n.d., p. 3, National Archives, AT 54/24.

50. J. Littlewood to T.M. Heiser, 'Play Space', 12 November 1971, National Archives, AT 54/24.

51. 'Special Report: From Five to Fourteen', *Play Times*, 6 (1978), 8–9.

52. Peter Smith, 'Time to Give Play a New Priority', *Municipal Review*, 44.519 (1973), 78–81 (p. 81).

53. National Playing Fields Association, 'DoE Circular 79/72 Children's Play Space: Note by the Children's Play Officer', 1972, National Archives, CB 1/61.

54. J.A. Goodburn to Mr Poore, 'Children's Play Space – NPFA Proposals', 23 November 1972, National Archives, HLG 118/1897.

55. Clare Cooper Marcus and Robin C. Moore, 'Children and Their Environments: A Review of Research 1955–1975', *Journal of Architectural Education*, 29.4 (1976), 22–5.

56. Geoffrey Hayward, Marilyn Rothenberg and Robert R. Beasley, 'Children's Play and Urban Playground Environments: A Comparison of Traditional, Contemporary, and Adventure Playground Types', *Environment and Behavior*, 6.2 (1974), 131–68.

57. Robin C. Moore and Donald Young, 'Childhood Outdoors: Toward a Social Ecology of the Landscape', in *Children and the Environment*, ed. Irwin Altman and Joachim F. Wohlwill, Human Behaviour and Environment: Advances in Theory and Research (New York: Plenum Press, 1978), III, 83–128 (p. 83).

58. Ivor Seeley, *Outdoor Recreation and the Urban Environment* (London: Palgrave Macmillan, 1973).

59. Iona Opie and Peter Opie, *Children's Games in Street and Playground* (Oxford: Clarendon Press, 1969), p. v.

60. Opie and Opie, *Children's Games*, pp. 12 and 16.

61. Ian Taylor and Paul Walton, 'Hey, Mister, This Is What We Really Do ...', in *Vandalism*, ed. Colin Ward (London: Architectural Press, 1973), pp. 91–5; Paul Thompson, 'The War with Adults', *Oral History*, 3.2 (1975), 29–38; Denis Wood, 'Free the Children! Down with Playgrounds!', *McGill Journal of Education*, 12.2 (1977), 227–42.

62. Colin Ward, 'Adventure Playground', *Freedom*, 6 September 1958, pp. 3–4; Colin Ward, 'Adventure Playground: A Parable of Anarchy', *Anarchy*, 7 (1961), 193–201; Colin Ward, *Anarchy in Action* (London: Allen & Unwin, 1973).

63. Colin Ward, *The Child in the Country* (London: Bedford Square Press, 1990), p. 10.

64. Colin Ward, *The Child in the City*, 2nd edn (London: Bedford Square Press, 1990), p. 73.

65. Ward, *The Child in the City*, p. 179.

66. Simon Nicholson, 'How NOT to Cheat Children: The Theory of Loose Parts', *Landscape Architecture*, 62.1 (1971), 30–34.

67. Simon Nicholson, 'The Theory of Loose Parts, an Important Principle for Design Methodology', *Studies in Design Education Craft & Technology*, 4.2 (1972), 5–14 (p. 6).

68. 'Ratepayers Petition Council for Children's Playground', *Western Times*, 24 April 1936, p. 7; 'More Playgrounds, Please, Say Parents', *Playing Fields Journal*, 12.1 (1952), 38.

69. '200 Children March on Town Hall', *The Guardian*, 27 August 1963, p. 12.

70. 'Children's Protest March', *The Guardian*, 13 August 1964, p. 14.

71. '200 Children "Lay Siege" to Town Hall', *The Guardian*, 5 September 1968, p. 18.

72. Greater London Council, 'Toddlers' Playgrounds: Play Provision on GLC Housing Estates', 1970, London Metropolitan Archive, GLC/HG/HHM/12/S026A.

73. *It's Ours Whatever They Say*, dir. by Jenny Barraclough, 1972, British Film Institute, BFI Player https://player.bfi.org.uk/free/film/watch-its-ours-whatever-they-say-1972-online [accessed 12 July 2023]. The site is now known as Biddestone Park.

74. *Do Something!*, dir. by Ross Devenish, 1970, British Film Institute, BFI Player https://player.bfi.org.uk/free/film/watch-do-something-1970-online [accessed 12 July 2023].

75. Pearl Jephcott, *A Troubled Area: Notes on Notting Hill* (London: Faber and Faber, 1964), p. 17.

76. Roger Mitton and Elizabeth Morrison, *A Community Project in Notting Dale* (London: Allen Lane, 1972); Jan O'Malley, *The Politics of Community Action: A Decade of Struggle in Notting Hill* (Nottingham: Spokesman Books, 1977); 'Children March through Melee', *Kensington Post*, 31 May 1968, p. 2; 'Children March to Town Hall', *Kensington Post*, 14 June 1968, p. 7.

77. *The Battle for Powis Square*, 1975, London Community Video Archive, CVA0019b https://lcva.gold.ac.uk/videos/5e5522d7c899b32870ce69d2 [accessed 12 July 2023].

78. National Playing Fields Association, 'Playground and Play Leadership Committee Minutes', 1963, National Archives, CB 1/55.

79. 'Drummond Retires', *Play Times*, 6 (1978), 3; Nick Balmforth, 'Drummond Abernethy OBE (1913–88)', *Journal of Playwork Practice*, 1 (2014), 101–3.

80. National Playing Fields Association, 'Institute of Playleadership Minutes, 9 February', 1970, National Archives, CB 1/64.

81. 'Out of the Way: Our Gang', *New Society*, 8 February 1973, pp. 305–6.

82. 'Politics and Play: Some Comments by a Play Worker', *Play Times*, 6 (1978), 2.

83. See, for example, Bob Hughes, *Notes for Adventure Playworkers* (London: Children and Youth Action Group, 1975) and subsequent university courses, academic journals and publications.

84. Nils Norman, *An Architecture of Play: A Survey of London's Adventure Playgrounds* (London: Four Corners Books, 2003).

85. National Playing Fields Association, 'Report of Adventure Playground Conference', 1956, National Archives, CB 1/67

86. Joe Benjamin, *Grounds for Play: An Extension of In Search of Adventure* (London: Bedford Square Press, 1974), p. 49.

87. John Barron Mays, *Adventure in Play: The Story of the Rathbone Street Adventure Playground* (Liverpool: Liverpool Council of Social Service, 1957).

88. Arvid Bengtsson, *Adventure Playgrounds* (London: Crosby Lockwood, 1972); Bernard McGovern, *Playleadership* (London: Faber and Faber, 1973); Jack Lambert and Jenny Pearson, *Adventure Playgrounds: A Personal Account of a Playleader's Work* (Harmondsworth: Penguin, 1974).

89. Joe Benjamin, 'Adventure for Industry', *New Society*, 13 December 1962, p. 22; Joe Benjamin, *In Search of Adventure* (London: National Council of Social Service, 1966); Benjamin, *Grounds for Play*.

90. Krista Cowman, '"The Atmosphere Is Permissive and Free": The Gendering of Activism in the British Adventure Playgrounds Movement, ca. 1948–70', *Journal of Social History*, 53.1 (2019), 218–41.

91. T.L. Cook, 'Children's Improvised Play Equipment', *Playing Fields Journal*, 30.2 (1970), 33–8.

92. Paul Hogan, *Playgrounds for Free* (Cambridge, MA: MIT Press, 1974); M. Paul Friedberg, *Do It Yourself Playgrounds* (London: Architectural Press, 1975); David Raphael, 'Grounds for Playful Renaissance', *Landscape Architecture*, 65.3 (1975), 329–30; 'DIY Playground Equipment', *Play Times*, 15 (1979), 15.

93. Bob Hughes, *Evolutionary Playwork*, 2nd edn (London: Routledge, 2013), p. 29.

94. J. Kennedy, 'Playground Planning', *Park Administration*, 32.8 (1967), 36–8.

95. 'Come out to Play in Battersea Park', *Park Administration, Horticulture and Recreation*, 27.5 (1962), 65.

96. London County Council, 'Play Parks', 1963, London Metropolitan Archives, LCC/PUB/11/01/119.

97. Marjory Allen, *Play Parks*, 3rd edn (London: Housing Centre, 1964).

98. 'Civic Trust Award', *Playing Fields Journal*, 30.2 (1970), 46; 'Children's Playgrounds', *L'Architecture d'Aujourd'hui*, 165 (1973), xxxv.

99. Ian C. Laurie, 'Public Parks and Spaces', in *Fifty Years of Landscape Design 1934–84*, ed. Sheila Harvey and Stephen Rettig (London: The Landscape Press, 1985), pp. 63–78 (p. 73).

100. Anne R. Beer, 'The External Environment of Housing Areas', *Built Environment*, 8.1 (1982), 25–9; Michael Haxeltine, 'A Check List for Play Spaces', *Parks and Recreation*, 38.6 (1973), 25–9; Lance H. Wuellner, 'Forty Guidelines for Playground Design', *Journal of Leisure Research*, 11.1 (1979), 4–14.

101. L. Potts, 'Children's Playgrounds', 21 May 1965, National Archives, Royal Parks, WORK 16/2299.

102. R.G. Emberson to Mr Barrow, 'Regent's Park – Gloucester Green Playground', 25 September 1969, National Archives, Royal Parks, WORK 16/2299; J.W. Gorvin to R.G. Emberson, 'Gloucester Green Children's Playground', 30 December 1970, National Archives, Royal Parks, WORK 16/2299.

103. Royal Parks, 'General Layout Plan for Gloucester Green Children's Playground', 1969, National Archives, WORK 16/2299.

104. 'International Landscape Awards – Gloucester Gate Playground by LUC', *Landezine*, 2021 https://landezine-award.com/gloucester-gate-playground/ [accessed 12 July 2023]; 'Donation Machines Next to Playground in Regent's Park', *Camden New Journal*, 26 March 2021.

105. 'Amalgamations', *Park Administration*, 31.11 (1966), 21; Philip Sayers, 'Whence and Whither? From Parks Superintendents to Leisure Planners', *Park Administration*, 31.12 (1966), 18–19.

106. National Playing Fields Association, 'Annual Report and Accounts', 1976, Museum of English Rural Life, SR CPRE C/1/73/2.

107. 'A Guide to Inter-Departmental Responsibilities for Children's Play', 1975, National Archives, AT 60/30.

108. 'Handwritten Note Regarding Guide to Inter-Departmental Responsibilities', 1979, National Archives, AT 60/30.

109. Otto Saumarez Smith, 'The Lost World of the British Leisure Centre', *History Workshop Journal*, 88 (2019), 180–203.

110. Wicksteed Village Trust, 'Wicksteed Park; 140 Acres of Fun and Freedom', 1970, Wicksteed Park Archive, BRC-1793; Wicksteed Village Trust, 'Hi! There's a Big Smile for Everyone to Wear at Wicksteed Park', n.d., Wicksteed Park Archive, BRC-1821.

111. Wicksteed Village Trust, 'Wicksteed Park Nature Trail', 1971, Wicksteed Park Archive, PRM-1883.

112. P.A. Management Consultants, 'Organisation and Control', 1967, Wicksteed Park Archive, uncatalogued.

113. Charles Wicksteed & Co., 'Design for Playing', *Playing Fields*, 25.4 (1965), 25.

114. 'Various Cuttings from Trade Journals and Newspapers', 1966, Wicksteed Park Archive, uncatalogued.

115. 'Six New Pieces of Playground Equipment', *Evening Telegraph*, 29 July 1964, p. 8.

116. Charles Wicksteed & Co., 'The Gateway to Health and Happiness through Wicksteed Playground Equipment', 1970, Wicksteed Park Archive, BRC-3010.

117. Charles Wicksteed & Co., 'When You Buy Wicksteed Equipment, You Receive Wicksteed Service', *Park Administration, Horticulture and Recreation*, 24.8 (1960), 392; H. Hunt & Son, 'Where It All Comes From', *Park Administration*, 28.1 (1963), 1.

118. H. Hunt & Son, 'Playground Equipment – the Verdict Is Yours', *Playing Fields*, 22.3 (1962), 9.

119. 'The Growing Variety in Playground Furniture', *The Municipal and Public Services Journal*, 75 (1967), 1272–3.

120. 'Triodetic Aluminium Playdome', *Park Administration*, 31.12 (1966), 39.

121. Playstyle Ltd, 'The Complete Playground: Playcubes', *Playing Fields*, 34.1 (1973), 10.

122. Sutcliffe Engineering, 'Seat for a Swing', 1978, Espacenet Patent Search, GB1535728A.

123. 'Catering for the Kids', *Parks and Recreation*, 40.6 (1975), 64–83; '"Intensive Use" Play Park', *Playing Fields*, 37.1 (1976), 64–5.

124. 'Conference Issue', *Parks and Recreation*, 42.6 (1977).

125. SMP Landscapes, 'Play Area Layout, Scheme B, Wicksteed Park Kettering', 1979, Wicksteed Park Archive, uncatalogued.

126. 'Children at Play', *Park Administration*, 34.6 (1969), 18–29; 'Keeping up to Date with Playground Equipment', *Parks and Recreation*, 42.10 (1977), 29–40.

127. 'Publications Available from the NPFA', *Playing Fields*, 37.4 (1976), 49–50.

128. 'Welcome to Your First Issue ...', *Play Times*, 1 (1977), 3.

129. Smith, 'Time to Give Play a New Priority', p. 78.

130. Newson and Newson, *Four Years Old in an Urban Community.*

131. Thomson, *Lost Freedom*, chapter 6.

132. Lucy Robinson, *Gay Men and the Left in Post-War Britain* (Manchester: Manchester University Press, 2007).

133. Neil Pemberton, 'The Burnley Dog War: The Politics of Dog-Walking and the Battle over Public Parks in Post-Industrial Britain', *Twentieth Century British History*, 28.2 (2017), 239–67.

134. J.A. Marron and C.L. Senn, 'Dog Feces: A Public Health and Environmental Problem', *Journal of Environmental Health*, 37 (1974), 239–43; A.W. Woodruff, 'Toxocariasis as a Public Health Problem', *Environmental Health*, 84 (1976), 29–31.

135. O.A. Borg and A.W. Woodruff, 'Prevalence of Infective Ova of Toxocara Species in Public Places', *British Medical Journal*, 4 (1973), 470.

136. Woodruff, 'Toxocariasis as a Public Health Problem', p. 29.

137. 'Pets Danger to Children Playing in Parks', *The Times*, 23 November 1973, p. 5; 'Why Do We Let Dogs Foul Our Streets?', *British Medical Journal*, 1 (1976), 1486; 'Girl Lost Sight of Eye from Disease Carried by Dogs', *The Times*, 25 June 1976, p. 2; 'Canine Menace', *The Sunday Times*, 3 April 1977, p. 13; 'Sting in the Tail: Man's Best Friend Can Also Be One of Children's Worst Enemies', *The Guardian*, 8 February 1989, p. 27.

138. W.G. Ayres, 'The Provision of Children's Playgrounds by a Local Authority', *Journal of Park Administration, Horticulture and Recreation*, 19.4 (1954), 151–60 (p. 153).

139. 'Accidents on Playgrounds: Local Authorities Discuss Liability Questions', *Playing Fields*, 14.3 (1954), 26; 'Play Chute Negligence Case', *Playing Fields*, 17.1 (1957), 48–9.

140. L.A. Huddart, 'Safety on the Playground', *Playing Fields*, 13.1 (1953), 42–50.

141. 'Developing the Rubberized Playground Surface', *Park Administration, Horticulture and Recreation*, 18.8 (1954), 342.

142. G.H. Harris, 'All-Rubber Playgrounds Reduce Accident Risks', *Playing Fields*, 14.4 (1954), 36–8.

143. London County Council, 'Unsupervised Play Space on Housing Estates', 1959, London Metropolitan Archives, GLC/HG/HHM/12/S026A.

144. London County Council, 'Playgrounds – Equipment', 1962, London Metropolitan Archives, GLC/HG/HHM/12/S026A.

145. London Standing Conference of Housing Estate Community Groups, 'Newsletter', 1960, London Metropolitan Archives, GLC/HG/HHM/12/S026A.

146. D.A. Routledge, R. Repetto-Wright and C.I. Howarth, 'The Exposure of Young Children to Accident Risk as Pedestrians', *Ergonomics*, 17.4 (1974), 457–80.

147. C. Illingworth and others, '200 Injuries Caused by Playground Equipment', *British Medical Journal*, 4.5992 (1975), 332 (p. 334).

148. Richard Carr, 'Playing Safe', *The Guardian*, 6 October 1972, p. 11.

149. 'Dangers for Children at Play', *The Times*, 6 June 1973, p. 10.

150. 'Park Safety Still Causing Concern', *Evening Telegraph*, 26 February 1968, Wicksteed Park Archive.

151. Richard Carr, 'Games of Chance', *The Guardian*, 23 April 1976, p. 9.

152. 'Playground Mishaps Injure 20,000', *The Guardian*, 6 May 1976, p. 6.

153. 'Peril of Playground Death Traps', *The Guardian*, 12 June 1978, p. 3.

154. 'Playground Probe Reveals Dangers in Swings and Slides', *The Sunday Times*, 4 June 1972, p. 3.

155. British Standards Institution, *B.S. 3178: Playground Equipment for Parks* (London: British Standards Institution, 1959); 'A British Standard for Children's Playground Equipment for Parks (Unbound Press Release Inserted into Journal)', *Park Administration, Horticulture and Recreation*, 24.6 (1959), 1–2.

156. Design Council, *Street Scene* (London: Design Council, 1976).

157. R.E.K. Holmes to M. Albu, 'Safety in Children's Playgrounds', 1977, National Archives, AT 54/159.

158. J. Littlewood to Mr. Oddy, 'Play Space Standards', 27 January 1978, National Archives, AT 54/159.

159. R.E.K. Holmes to Local Authority Chief Executives, 'The Need for Improved Safety in Children's Playgrounds', 31 October 1978, National Archives, AT 54/159.

160. Fair Play for Children, 'Press Statement', 3 November 1978, National Archives, AT 54/159; Bob Satterthwaite to R.E.K. Holmes, 'The Need for Improved Safety in Children's Playgrounds', 24 November 1978, National Archives, AT 54/159.

161. D.F. Hodson to R.E.K. Holmes, 'Borough of Thamesdown Response to DoE Letter on Children's Playgrounds', 4 December 1978, p. 1, National Archives, AT 54/159.

162. R.E. Gamble to Mrs. Watson, 'Playground Safety', 7 August 1978, National Archives, AT 54/159.

163. Benjamin, *Grounds for Play*, p. 51.

164. Harry Shier, *Adventure Playgrounds: An Introduction* (London: National Playing Fields Association, 1984).

165. 'Northumberland Road Adventure Playground and the Health and Safety at Work Act, 1974', *Play Times*, 14 (1979), 2.

166. 'Health and Safety Act', *Play Times*, 15 (1979), 2.

167. 'Helping with Safety in Manchester', *Play Times*, 16 (1979), 2.

168. Hughes, *Evolutionary Playwork*, p. 29.

169. National Playing Fields Association, *Towards a Safer Adventure Playground* (London: National Playing Fields Association, 1980).

170. 'Factors in the Development of the Playing Field Movement', *Playing Fields Journal*, 19.1 (1959), 25–6; 'Reductions in Capital Expenditure from Public Funds', *Playing Fields*, 25.4 (1965), 26.

171. Stewart Harding, 'Towards a Renaissance in Urban Parks', *Cultural Trends*, 9.35 (1999), 1–20.

172. Liz Greenhalgh and Ken Worpole, *Park Life: Urban Parks and Social Renewal* (London: Comedia/Demos, 1995); David Lambert, *The Park Keeper* (London: English Heritage, 2005).

173. Katy Layton-Jones and Robert Lee, *Places of Health and Amusement: Liverpool's Historic Parks and Gardens* (Swindon: English Heritage, 2008).

174. B. Clouston, 'Urban Parks in Crisis', *Landscape Design*, 149, 1984, 12–14.

175. 'Safety on Play Areas', *Play Times*, 1 (1977), 11.

176. 'The X-Certificate Playgrounds', *Play Times*, 2 (1977), 6–7.

177. 'Brick of the Month Award – Portsmouth', *Play Times*, 14 (1979), 4; 'Brick of the Month – Neath', *Play Times*, 17 (1979), 13.

178. 'Brick of the Month – Knowsley', *Play Times*, 16 (1979), 16.

179. Alison Ravetz, *Model Estate: Planned Housing at Quarry Hill, Leeds* (London: Croom Helm, 1974), pp. 141–50.

180. Nigel Lawson to Michael Heseltine, 'Housing Subsidy System and Project Control', 7 July 1980, National Archives, HLG 118/1897; Mr. Stanley, *House of Commons Debate, 19 January 1981, Vol.997, Col.66, Play Space* (Hansard, 1981).

181. Dennis Barker, 'Self-Help Is Child's Play for Thatcher', *The Guardian*, 12 July 1980, p. 3.

182. Ian Nicol to Mr Moss, 'New Subsidy Arrangements: Playspace Provision', 6 October 1980, National Archives, HLG 118/1897.

183. R.V.G. Clarke, ed., *Tackling Vandalism* Home Office Research Study, 47 (London: HMSO, 1978).

184. Alice Coleman, *Utopia on Trial: Vision and Reality in Planned Housing* (London: Hilary Shipman, 1985), p. 47.

185. Coleman, *Utopia on Trial*, p. 78.

186. Peter Dickens, 'Utopia on Trial: A Response to Alice Coleman's Comment', *International Journal of Urban and Regional Research*, 11 (1987), 118–20; Paul Spicker, 'Poverty and Depressed Estates: A Critique of Utopia on Trial', *Housing Studies*, 2.4 (1987), 283–92.

187. Christopher White, 'Rampage of the Tiny Vandals, Aged 10', *Daily Mail*, 30 April 1985, p. 1; 'Vandals Leave £1.8bn Bill and a Trail of Fear', *The Sunday Times*, 13 August 1989, p. 5; Geoffrey Levy, 'Playground Monsters', *Daily Mail*, 16 May 1990, p. 6; David Charter, 'The War in the Playground', *The Times*, 29 April 1997, p. 20.

188. 'BSI Standards for Playground Equipment', *Parks and Recreation*, 44.3 (1979), 30–31; 'For Safety's Sake: The New British Standard on Playground Equipment', *Play Times*, 13 (1979), 11.

189. British Standards Institution, *B.S. 5696: Play Equipment Intended for Permanent Installation Outdoors* (London: British Standards Institution, 1979).

190. 'A Surface Guide for Children's Playgrounds and Multi-Purpose Play Areas', *Play Times*, 5 (1977), 11.

191. Peter Jones, 'Re-Thinking Corruption in Post-1950 Urban Britain: The Poulson Affair, 1972–1976', *Urban History*, 39.3 (2012), 510–28.

192. L.B. Creasey, 'A Visit to Wicksteed's of Kettering', *Park Administration, Horticulture and Recreation*, 22 (1958), 632.

193. 'A New Type of Play Area', *Park Administration, Horticulture and Recreation*, 26.1 (1961), 59.

194. Guy Ortolano, *Thatcher's Progress: From Social Democracy to Market Liberalism through an English New Town* (Cambridge: Cambridge University Press, 2019), p. 257.

195. John McKendrick, Anna Fielder and Michael Bradford, 'Privatization of Collective Play Spaces in the UK', *Built Environment*, 25.1 (1999), 44–57.

196. Susan Markwell and Neil Ravenscroft, *The Public Provision of Children's Playgrounds* (Reading: University of Reading, 1996).

197. Peter Heseltine, *The Children's Playground: A Basic Guide* (London: Royal Society for the Prevention of Accidents, 1997); Peter Heseltine, *Regular Inspection of Children's Playgrounds* (London: Royal Society for the Prevention of Accidents, 1999).

198. John McKendrick, 'Playgrounds in the Built Environment', *Built Environment*, 25.1 (1999), 5–10.

199. National Playing Fields Association, *Best Play: What Play Provision Should Do for Children* (London: National Playing Fields Association, 2000).

200. Play England, *Developing an Adventure Playground: The Essential Elements*, Practice Briefing (London, 2009).

201. Tim Gill, *No Fear: Growing Up in a Risk Averse Society* (London: Calouste Gulbenkian Foundation, 2007); Adrian Voce, *Policy for Play: Responding to Children's Forgotten Right* (Bristol: Bristol University Press, 2015).

202. Welsh Assembly Government, *Play Policy* (Cardiff: Welsh Assembly, 2002); The Scottish Government, *Play Strategy for Scotland: Our Vision* (Edinburgh: Scottish Government, 2013).

203. Aileen Shackell and others, *Design for Play: A Guide to Creating Successful Play Spaces* (London: Department for Children, Schools and Families and Department for Culture, Media and Sport, 2008).

204. Richard Adams, 'Hundreds of Children's Playgrounds in England Close Due to Cuts', *The Guardian*, 13 April 2017; Harriet Grant, Aamna Mohdin and Chris Michael, '"Outrageous" and "Disgusting": Segregated Playground Sparks Fury', *The Guardian*, 26 March 2019.

205. Jon Winder, 'Children's Playgrounds: "Inadequacies and Mediocrities Inherited from the Past"?', *Children's Geographies*, 2023, 1–6 https://doi.org/10.1080/14733285.2023.2197577.

206. 'Future Role of the NPFA', 1970, National Archives, HLG 120/1614.

207. Robin C. Moore, 'Playgrounds at the Crossroads', in *Public Places and Spaces*, ed. Irwin Altman and Ervin H. Zube (Boston, MA: Springer, 1989), pp. 83–120.

Conclusion

The children's playground is an everyday feature of British towns and cities. Often seen as the obvious place for children to play, these apparently simple spaces of pleasure are the tangible expression of significant and interconnected historical themes. In exploring playground form and function over the course of two centuries, *Designed for Play* has provided important new perspectives on the histories and geographies of childhood, nature, welfare, philanthropy, education and the urban environment. The book has tracked the ideas and practical actions that have sought to channel childhood playfulness, exposing an enduring tension between a universal urge to play and adult attempts to create public spaces where this should take place. A detailed analysis of dispersed archive material has provided a more nuanced and multilayered understanding of the playground, the key actors involved in its development and its social, political and environmental rationales. In plotting these processes for the first time, the book has shown how a diverse set of assumptions about human wellbeing – pursued by state, philanthropic, commercial and voluntary actors – has had a lasting influence on the material form of the playground and the politics of urban space. But the evolution of the playground has been far from linear or straightforward. At times a public plantation focused on nurturing young minds and muscles, the playground has also been associated more disparagingly with monstrous gangs of troublesome children.

As such, the fortunes of the playground have swung back and forth, much like the motion of a swing, from marginal obscurity to popular ubiquity and back again, towards a place of somewhat aimless eccentricity. Similarly, many of the wider themes that have shaped playground discourse and practice have varied in importance over time, often coming full circle, much like the rotation of a roundabout. In particular, assumptions about the restorative potential of nature have consistently orbited the playground ideal. At times passing close by, nature provided a central justification for the children's garden gymnasiums of the 1890s and inspiration for landscape architects in the 1960s. At other times, nature's trajectory took it to the margins of the playground, and it was barely visible in the 1930s orthodox playground or in anxieties about safety in the late twentieth century. In contrast, conceptions of the playground as a place of health have tracked less obviously onto the cyclical movements of manufactured equipment. Instead, dedicated public spaces for children were consistently positioned as sites of salubrious safety from the mid-nineteenth century through to the late twentieth century. Initially, energetic exercise was understood as a vector for physical strength and vigour, while progressive education would later help to shape the playground as a site of emotional health and cognitive development. In practice, playgrounds had long posed risks to children and by the 1970s these threats were increasingly seen to outweigh the benefits that dedicated play spaces could provide. Technology, particularly in the form of manufactured equipment, has provided a lasting influence on the material form of public playgrounds, despite highly critical and enduring condemnation. Initially expressed in the 1930s but seen most notably in mid-twentieth-century anarchic thought and radical experiments in adventurous play, this criticism continues to reverberate today.

For the first time, the chapters in this book have plotted the interaction of these processes in more detail over time and space, positioning the children's playground as a site where social, political and environmental values have long been played out and contested. From the sporadic attempts to create dedicated public spaces for children in the mid-nineteenth century to the work of the Metropolitan Public Gardens Association in the 1890s, it has explored the links between conceptions of the city, nature, childhood and health. Charles Dickens's unsuccessful Playground Society showed how the provision of dedicated spaces for play would require a diverse set of ideas, values and assumptions to coalesce before material change could take place. Later in the nineteenth century, social and political anxiety about both cities and childhood would combine with voluntary action and philanthropy to create smaller, more local public gardens with a focus on energetic physical exercise and interaction with nature. However, while the principle of creating dedicated places for children's recreation became more firmly established at the turn of the century, the form of such spaces was far from settled.

In the early twentieth century, one vision for the playground increasingly acquired 'orthodox' status, based in part on the ideas, products and playground promoted by Charles Wicksteed. Inspired by amusement park rides and progressive attitudes to childhood and education, the playground was reimagined as a site of excitement for all, and increasingly featured in visions for modern, planned urban environments. As municipal officials, architects and planners delivered new city spaces, the number of public playgrounds increased significantly during the interwar years, in part a response to the active campaigning of the National Playing Fields Association. At the same time, the swing and other manufactured equipment came to dominate playground spaces, to the exclusion of naturalistic features. The principle of the playground indirectly benefited from the wider mid-century welfare consensus and its emphasis on the wellbeing of children and their families. At the same time, this focus also provided the foundations for an increasingly critical reception for the orthodox playground, as a diverse range of activists and practitioners emphasised children's self-determination, adventure and creativity.

During the 1970s, a wave of sociological research, anarchic thought, community activism and wider attempts to promote child liberation all challenged conventional playground thought. A backlash against these wider values from the 1980s saw anxieties about pets and paedophiles contribute to an assessment of the playground as a problem to be solved, rather than a place of promise and potential. While a belief in the playground as a space of safety had long concealed possible threats, when wider social discourse created an atmosphere in which these threats seemed increasingly insurmountable, perceptions of the playground shifted among park managers, politicians and the public. Exacerbated by cuts to local authority budgets, from the late 1970s through to the early twenty-first century, such spaces were increasingly perceived as spaces of danger and decay, inadvertently supporting radical attempts to undermine the case for playground provision altogether.

Designed for Play has identified the main actors involved in shaping playground provision and explored their assumptions and motivations over the course of two centuries. In doing so, the book has shown how the term 'playground' has proven sufficiently flexible to accommodate many revisions to its meaning, while largely retaining its core association with spaces of purposeful and healthy recreation. The narration of this story has placed children more centrally in our understanding of the nineteenth-century public parks movement, initially making sense of their absence from such spaces, before examining their increasing presence from the 1880s onwards. Assumptions about urban nature were significant in such processes, as were changing attitudes towards park-based recreation, notably the shift in emphasis from genteel perambulation to more energetic exercise. Spaces for play have long embodied and reflected wider social norms, including segregation by gender in the late nineteenth century, an emphasis on popular leisure activities in the interwar period, and the ongoing place of philanthropy and voluntary action in shaping both public life and public space.

The case study of Charles Wicksteed, his company and the playground he created provides a comprehensive account of this highly significant playground advocate and charts his enduring influence on international visions for the playground. He combined progressive notions of childhood, industrial philanthropy and a direct connection to the ideas of other garden city advocates, cementing the playground's place in modern planned visions for housing and the wider urban landscape. The book has built upon existing scholarship that charts the transatlantic exchange of park ideals, but also goes further to highlight the wider twentieth-century connections with playground thought in Europe too. In doing so, it acknowledges international influences on playground provision, but also examines the specific social and cultural factors that shaped public play space provision in Britain. As a result, it extends existing scholarship on the mid-twentieth-century adventure playground by moving beyond the boundary fence to assess the influence of such spaces on wider playground provision.

The playground has served as a valuable site for exploring wider historical themes and their previously unacknowledged influence on the built form of towns and cities. *Designed for Play* has shown how the radical visionaries who reimagined and redefined spaces for play in the city were often women. From Fanny Wilkinson and Mabel Jane Reaney to Marjory Allen, Mary Mitchell and Margaret Willis, the book has drawn attention to a succession of pioneering advocates and designers who challenged established ideas about children's place in the urban environment and shaped alternative spaces for play. Given the enduring influence of their ideas and actions, a more detailed and critical engagement with their work could usefully inform present-day calls to create more just and equitable cities.

Beyond the work of key individuals, the story of the playground provides a unique example of the long evolution of welfare interventions in the public realm and the varied outcomes they were expected to achieve. Importantly, the history of the playground demonstrates how a diverse set of actors across the philanthropic, voluntary, state and commercial sectors all sought to reimagine and reshape the urban landscape to improve childhood outcomes. Tentative efforts to introduce playgrounds as a route to childhood health in the late nineteenth and early twentieth century expanded considerably in the interwar and postwar years. The playground ideal was flexible enough to form part of both municipal leisure provision in the 1920s and 1930s and mid-century social democratic welfare landscapes too. As a result, playground provision expanded substantially over the course of the twentieth century; in Edinburgh, for example, the city's 15 children's gymnasiums in 1914 had increased to over 160 children's playgrounds by the early twenty-first century. At the same time, the enduring involvement of commercial equipment suppliers in shaping the form and function of public spaces for play undoubtedly complicates the place of the playground in narratives of a later twentieth-century shift towards market liberalism.

This new understanding of the history of the playground raises important questions for researchers, policymakers and practitioners. In adopting a long chronology, *Designed for Play* has focused less on the detailed stories of individual sites and the local political and cultural values that shaped their design and use. Where source materials have allowed, the ways in which children adapted and contested adult expectations of the playground have been stressed. Although sensitive to addressing children's position as 'academic orphans' in the field, uncovering their hidden voices has barely been possible and it undoubtedly warrants further exploration. In addition, further research into the global spread of the orthodox playground ideal in the interwar period would usefully inform discussion about Britain's relationship with its colonial past. *Designed for Play* has shown how play spaces in Britain were influenced by international ideas from Europe and north America and how the orthodox playground ideal and items of equipment were exported by Wicksteed & Co. However, there remains considerable scope to explore the themes uncovered here in other places around the world and archival material in Cape Town and Johannesburg remains ripe for investigation. In twenty-first-century Britain, the politics of the playground are far from settled. The Make Space for Girls campaign and Playing Out movement are rightly challenging male-dominated, car-centric approaches to public space design. Pay-to-play playgrounds in Windsor Great Park and Alnwick Gardens, with their £16 per child entry fees, raise significant questions about who can afford to access spaces for play. For policymakers and campaigners, the historical context outlined here encourages deeper reflection on the values and assumptions that shape children's place in public space and provides a new starting point for conversations with their communities, political representatives and funders.

Designed for Play has shown that the children's playground has long been a site where adult anxiety about public childhood has been played out. The form and function of such spaces have changed over time in response to shifting ideas about the benefits of interaction with nature, energetic exercise, entertainment and adventure. Enthusiasm for the principle of the playground has similarly fluctuated in response to changing conceptions of childhood, philanthropic funding, notions of safety and the power of utopian visions for better cities. Present-day playgrounds, along with their advocates and detractors, continue to embody and question the significance of these themes. Similarly, seemingly novel calls to re-wild childhood, reintroduce nature into cities and re-energise children through physical exercise are not new concerns; they too have a long and significant history. Situating twenty-first-century efforts to create more equitable and inclusive urban environments within the historical context outlined here will contribute to more constructive dialogue about children's place in the city.

References

Primary Sources

'200 Children "Lay Siege" to Town Hall', *The Guardian*, 5 September 1968, p. 18
'200 Children March on Town Hall', *The Guardian*, 27 August 1963, p. 12
'A British Standard for Children's Playground Equipment for Parks (Unbound Press Release Inserted into Journal)', *Park Administration, Horticulture and Recreation*, 24.6 (1959), 1–2
A Few Pages about Manchester (Manchester: Love and Barton, 1850)
'A Great Day for Paddington: Montgomery of Alamein Opens Rebuilt Recreation Ground', *Journal of Park Administration, Horticulture and Recreation*, 13.1 (1948), 29–30
'A Guide to Inter-Departmental Responsibilities for Children's Play', 1975, National Archives, AT 60/30
'A New Type of Play Area', *Park Administration, Horticulture and Recreation*, 26.1 (1961), 59
'A Recreation Ground Swing Accident', *Playing Fields Journal*, 2.6 (1934), 266
'A Surface Guide for Children's Playgrounds and Multi-Purpose Play Areas', *Play Times*, 5 (1977), 11
'A Want of the Age', *Bell's Life in London*, 17 January 1858, p. 7
Abercrombie, Patrick, and J.H. Forshaw, *County of London Plan* (London: Macmillan, 1943)
Abernethy, W.D., 'Children's Playgrounds', 1968, National Archives, AT 54/24
Abernethy, W.D., 'Report on United Nations Playground Seminar', 1958, National Archives, National Playing Fields Association CB 1/70
Abernethy, W.D., 'What Play Leadership Implies', *Park Administration*, 30.3 (1965), 55
'Accidents on Playgrounds: Local Authorities Discuss Liability Questions', *Playing Fields*, 14.3 (1954), 26
Adams, Richard, 'Hundreds of Children's Playgrounds in England Close Due to Cuts', *The Guardian*, 13 April 2017
Adams, Thomas, *Outline of Town and City Planning: A Review of Past Efforts and Modern Aims* (London: J & A Churchill, 1936)

Adams, Thomas, *PlayParks with Suggestions for Their Design, Equipment and Planting* (London: Coronation Planting Committee, 1937)
Addis, Ian, *Out to Play in Kettering* (Kettering: Bowden Publications, 2013)
'Adventure Playground for Handicapped Children', 1970, Wellcome Collection, Robina Addis Archives PP/ADD/K/2/2
'Advert for Wicksteed & Co. Playground Equipment', *Journal of Park Administration, Horticulture and Recreation*, 1.1 (1936), 1
'Advertisements', *La Belle Assemblée*, 1 July 1825
Agar, Madeline, *Garden Design in Theory and Practice*, 2nd edn (London: Sidgwick & Jackson, 1913)
'Albert Park', 1851, National Archives, WORK 32/424
Aldis, Philip, 'Churchill Gardens', *New Left Review*, 10 (1961), 55–9
Allen, Marjory, *Adventure Playgrounds* (London: National Playing Fields Association, 1953)
Allen, Marjory, 'Children in "Homes"', *The Times*, 15 July 1944, p. 5
Allen, Marjory, *Design for Play: The Youngest Children* (London: Housing Centre, 1962)
Allen, Marjory, 'Juvenile Delinquency', *The Times*, 6 December 1948, p. 5
Allen, Marjory, 'Letter to the Editor: Children's Playgrounds', *The Times*, 12 December 1952, p. 9
Allen, Marjory, *New Houses, New Schools, New Citizens* (London: Nursery School Association of Great Britain, 1934)
Allen, Marjory, *Planning for Play* (London: Thames and Hudson, 1968)
Allen, Marjory, *Play Parks*, 3rd edn (London: Housing Centre, 1964)
Allen, Marjory, 'The Coronation and the Village', *The Spectator*, 12 March 1937, p. 467
Allen, Marjory, *Whose Children?* (London: Simpkin Marshall, 1945)
Allen, Marjory, 'Why Not Use Our Bomb Sites like This?', *Picture Post*, 16 November 1946, 26–9
Allen, Marjory, 'Why the Stockholm Playgrounds Are So Successful', *The Architect and Building News*, 30 December 1954, pp. 812–14
Allen, Marjory, and Mary Nicholson, *Memoirs of an Uneducated Lady: Lady Allen of Hurtwood* (London: Thames and Hudson, 1975)
'Amalgamations', *Park Administration*, 31.11 (1966), 21
Amherst, Alicia, *London Parks and Gardens* (London: Archibald Constable & Co., 1907)
'An Imaginative Approach to Playground Provision', *Park Administration*, 28.12 (1963), 42
'An Open Space', *The Guardian*, 26 April 1940, p. 4
'An Organized Playground', *The Times*, 8 July 1909, p. 9

Anderson, Sir John. Letter to Lawrence Chubb, Letter, 25 April 1944, National Archives, CB 1/76
'Are Our Child Exiles Happy? A Plea for Play in Reception Areas', *Journal of Park Administration, Horticulture and Recreation*, 4.9 (1940), 213–15
'ASLA Notes', *Landscape Architecture*, 21.2 (1931), 139–45
Astor, Nancy, *House of Commons Debate, 28 April 1926, Vol.194, Col.2155* (Hansard, 1926)
Ayres, W.G., 'The Provision of Children's Playgrounds by a Local Authority', *Journal of Park Administration, Horticulture and Recreation*, 19.4 (1954), 151–60
Banham, Reyner, Paul Barker, Peter Hall and Cedric Price, 'Non-Plan: An Experiment in Freedom', *New Society*, 20 March 1969, pp. 435–43
Barker, Dennis, 'Self-Help Is Child's Play for Thatcher', *The Guardian*, 12 July 1980, p. 3
Barnes, Alfred, *House of Commons Debate, 28 March 1950, Vol.473, Col.33* (Hansard, 1950)
Bayliss, Jones and Bayliss Ltd, 'Gymnasia for Parks and Recreation Grounds, School Playgrounds, Etc.', 1912, National Archives, WORKS/16/1705
Beames, Thomas, *The Rookeries of London: Past, Present and Prospective* (London: Thomas Bosworth, 1850)
Beer, Anne R., 'The External Environment of Housing Areas', *Built Environment*, 8.1 (1982), 25–9
Bengtsson, Arvid, *Adventure Playgrounds* (London: Crosby Lockwood, 1972)
Bengtsson, Arvid, 'Children's Playground in a Swedish Town', *Park Administration, Horticulture and Recreation*, 22.10 (1958), 478–9
Bengtsson, Arvid, *Environmental Planning for Children's Play* (London: Lockwood, 1970)
Benjamin, Joe, 'Adventure for Industry', *New Society*, 13 December 1962, p. 22
Benjamin, Joe, *Grounds for Play: An Extension of In Search of Adventure* (London: Bedford Square Press, 1974)
Benjamin, Joe, *In Search of Adventure* (London: National Council of Social Service, 1966)
Bentley, Jane, and Laura Freeman, 'Play Space: The Design, Research and Development of a Play Area for Courtenay Special School, Stoke Lyne Hospital, Exmouth', 1972, Museum of English Rural Life, Landscape Institute, Pamphlet 2870 Box 1/08

Bergmann, J., 'The Parks of Copenhagen', *Journal of Park Administration, Horticulture and Recreation*, 13.1 (1948), 18–21
'Bermondsey Children's "Variety" Playground', *Playing Fields Journal*, 12.1 (1952), 35–7
Besant, Walter, 'The Social Wants of London: IV Gardens and Playgrounds', *The Pall Mall Gazette*, 18 March 1884, pp. 1–2
Blowers, A.T., 'Council Housing: The Social Implications of Layout and Design in an Urban Fringe Estate', *The Town Planning Review*, 41.1 (1970), 80–92
Board of Education and Ministry of Transport, *Report of the Inter-Departmental Committee (England & Wales) on Road Safety Among School Children* (London: HMSO, 1936)
'Boarding School for Young Gentleman at Ilford in Essex', *Morning Chronicle*, 1 August 1795, p. 5
Boddy, F.A., 'Playgrounds for Children', *Journal of Park Administration, Horticulture and Recreation*, 3.11 (1939), 345
'Bomb on East End School', *The Times*, 14 June 1917, p. 8
Booth, William, *In Darkest England and the Way Out* (London: Salvation Army, 1890)
Borg, O.A., and A.W. Woodruff, 'Prevalence of Infective Ova of Toxocara Species in Public Places', *British Medical Journal*, 4 (1973), 470
Brabazon, Reginald, *Social Arrows*, 2nd edn (London: Longmans, Green & Co., 1887)
Bray, Reginald, 'The Children of the Town', in *The Heart of Empire: Discussions of Problems of Modern City Life in England, with an Essay on Imperialism*, edited by Charles Masterman (London: Fisher Unwin, 1902), pp. 111–64
'Breathing Space in London', *Park Administration, Horticulture and Recreation*, 24.12 (1960), 667–9
'Brick of the Month – Knowsley', *Play Times*, 16 (1979), 16
'Brick of the Month – Neath', *Play Times*, 17 (1979), 13
'Brick of the Month Award – Portsmouth', *Play Times*, 14 (1979), 4
'Brighter London Parks: Mr. Lansbury Continues His Tour', *The Manchester Guardian*, 9 October 1929, p. 17
British Standards Institution, *B.S. 3178: Playground Equipment for Parks* (London: British Standards Institution, 1959)
British Standards Institution, *B.S. 5696: Play Equipment Intended for Permanent Installation Outdoors* (London: British Standards Institution, 1979)
Brown, Michael, 'Drawings and Plans in the Michael Brown Collection', 1966, Museum of English Rural Life, AR BRO DO

Brown, Michael, ‘Landscape and Housing’, *Official Architecture and Planning*, 30.6 (1967), 791–9
Bruner, J.S., A. Jolly and K. Sylva, eds., *Play: Its Role in Development and Evolution* (Harmondsworth: Penguin, 1976)
‘BSI Standards for Playground Equipment’, *Parks and Recreation*, 44.3 (1979), 30–31
Bush, F.R., Letter to Chief Commissioner of Police, letter, 13 February 1930, National Archives, MEPO 2/7803
Campbell, Janet M., *The Physical Welfare of Mothers and Children* (Dunfermline: Carnegie United Kingdom Trust, 1917)
‘Canine Menace’, *The Sunday Times*, 3 April 1977, p. 13
Cantell, Timothy, *Urban Wasteland: A Report on Land Lying Dormant in Cities, Towns and Villages in Britain* (London: Civic Trust, 1977)
Carr, Richard, ‘Games of Chance’, *The Guardian*, 23 April 1976, p. 9
Carr, Richard, ‘Playing Safe’, *The Guardian*, 6 October 1972, p. 11
‘Catering for the Kids’, *Parks and Recreation*, 40.6 (1975), 64–83
Chandos Leigh, E., ‘The London Playing Fields Society’, *The Times*, 20 January 1911, p. 19
Charles Wicksteed & Co., ‘Design for Playing’, *Playing Fields*, 25.4 (1965), 25
Charles Wicksteed & Co., ‘Director Minute Book, 1920–1956’, Wicksteed Park Archive, uncatalogued
Charles Wicksteed & Co., Letter to City of Lincoln Surveyor, ‘Children’s Playground Equipment’, 21 November 1933, Wicksteed Park Archive, uncatalogued
Charles Wicksteed & Co., ‘Play Things as Used in the Wicksteed Park’, 1923, National Archives, WORKS/16/1705
Charles Wicksteed & Co., ‘Playground Equipment, Tennis Posts, Fencing and Park Seats’, 1926, Wicksteed Park Archive, uncatalogued
Charles Wicksteed & Co., ‘Price List for Playground Equipment’, 1949, London Metropolitan Archive, GLC/RA/D2G/04/091
Charles Wicksteed & Co., ‘The Gateway to Health and Happiness through Wicksteed Playground Equipment’, 1970, Wicksteed Park Archive, BRC-3010
Charles Wicksteed & Co., ‘When You Buy Wicksteed Equipment, You Receive Wicksteed Service’, *Park Administration, Horticulture and Recreation*, 24.8 (1960), 392
Charter, David, ‘The War in the Playground’, *The Times*, 29 April 1997, p. 20
Chawner, Thomas, and James Pennethorne, ‘Plan for Laying out the Proposed Eastern Park to Be Called Victoria Park’, 1841, British Library, Maps.Crace XIX 43a

'Children and Planning', *Town and Country Planning*, 36.10–11 (1968), 430–512
'Children at Play', *Park Administration*, 34.6 (1969), 18–29
'Children in Kensington Gardens', *The Times*, 7 August 1909, p. 6
'Children in Towns', *The Times*, 11 March 1944, p. 5
'Children March through Melee', *Kensington Post*, 31 May 1968, p. 2
'Children March to Town Hall', *Kensington Post*, 14 June 1968, p. 7
'Children Revel in Street Playground', *New York Times*, 26 July 1914, p. 11
'Children's Corner', *Playing Fields Journal*, 1.8 (1931), 29
'Children's Playground in South London', *Illustrated London News*, 10 May 1884, p. 443
'Children's Playgrounds', *L'Architecture d'Aujourd'hui*, 165 (1973), xxxv
'Children's Protest March', *The Guardian*, 13 August 1964, p. 14
'Child's Injury in Playground: Damages Doubled, Corporation Loses Appeal', *The Manchester Guardian*, 14 July 1937, p. 15
Chiswick Polytechnic, 'Hospital Play Specialist Course', 1973, National Archives, MH 152/134
Chubb, Lawrence, Letter to Ernest Holderness, 'Street Closures for Children's Play', 17 October 1932, National Archives, MEPO 2/7803
'Civic Trust Award', *Playing Fields Journal*, 30.2 (1970), 46
Clark, H.F., 'A New Type of Park Administrator', *Park Administration, Horticulture and Recreation*, 23.11 (1959), 439–41
Clarke, R.V.G., ed., *Tackling Vandalism*, Home Office Research Study, 47 (London: HMSO, 1978)
Cleeve Barr, A.W., *Public Authority Housing* (London: Batsford, 1958)
Clias, Peter Henry, *An Elementary Course of Gymnastic Exercises* (London: Sherwood, Jones & Co., 1823)
Clouston, B., 'Urban Parks in Crisis', *Landscape Design*, 149 (1984), 12–14
Coates, William, 'The Duty of the Medical Profession in the Prevention of National Deterioration', *British Medical Journal*, 1 (1909), 1045–50
Coleman, Alice, *Utopia on Trial: Vision and Reality in Planned Housing* (London: Hilary Shipman, 1985)
Colvin, Brenda, *Land and Landscape* (London: John Murray, 1948)
Come out to Play, 1954, British Pathé Archive, DOCS 1359.01 https://www.britishpathe.com/video/come-out-to-play-reel-1-1/ [accessed 6 July 2023]
'Come out to Play in Battersea Park', *Park Administration, Horticulture and Recreation*, 27.5 (1962), 65
'Conference Issue', *Parks and Recreation*, 42.6 (1977), 1–106
Cook, T.L., 'Children's Improvised Play Equipment', *Playing Fields Journal*, 30.2 (1970), 33–8

Cooper Marcus, Clare, and Robin C. Moore, 'Children and Their Environments: A Review of Research 1955–1975', *Journal of Architectural Education*, 29.4 (1976), 22–5
Coote, B.T., 'Children's Playgrounds: Their Equipment and Use', *Journal of Park Administration, Horticulture and Recreation*, 1.2 (1936), 102–5
Coronation Planting Committee, *The Royal Record of Tree Planting, the Provision of Open Spaces, Recreation Grounds & Other Schemes Undertaken in the British Empire and Elsewhere, Especially in the United States of America, in Honour of the Coronation of His Majesty King George VI* (Cambridge: Cambridge University Press, 1939)
'Corporations and Public Playgrounds: A Question of Liability', *The Manchester Guardian*, 10 January 1905, p. 12
'Court of Appeal', *The Times*, 15 March 1934, p. 4
Cowes, Dudley S., 'Leave This to Us, Sonny – You Ought to Be out of London', n.d., Imperial War Museum, Art.IWM PST 15093
Crane, Susan A., 'Historical Subjectivity: A Review Essay', *The Journal of Modern History*, 78.2 (2006), 434–56
Creasey, L.B., 'A Visit to Wicksteed's of Kettering', *Park Administration, Horticulture and Recreation*, 22 (1958), 632
'Crown Court: £1,500 Damages for Boy, Playground Accident', *The Manchester Guardian*, 22 December 1936, p. 5
Cruickshank, Marjorie, 'The Open-Air School Movement in English Education', *Paedagogica Historica*, 17.1 (1977), 62–74
Curtis, Henry S., *Education Through Play* (New York: Macmillan, 1915)
Curtis-Bennett, Noel, 'Playing Fields in the Post-War Period', *Journal of Park Administration, Horticulture and Recreation*, 10.3 (1945), 47–63
'Dangers for Children at Play', *The Times*, 6 June 1973, p. 10
De Morgan, Sophia Elizabeth, *Memoir of Augustus De Morgan* (London: Longmans Green, 1882)
Denby, Elizabeth, *Europe Re-Housed* (London: Allen & Unwin, 1938)
Denby, Elizabeth, 'Rehousing from the Slum Dweller's Point of View', *Journal of the Royal Institute of British Architects*, 44 (1936), 61–80
Department of Health and Social Security, *Play for Children in Hospital*, Circular HC(76)5 (London: HMSO, 1976)
Department of the Environment, *Children at Play*, Design Bulletin, 27 (London: HMSO, 1973)
Department of the Environment and Welsh Office, *Children's Playspace*, Circular 79/72 and 165/72 (London: HMSO, 1972)
Deputy Education Officer, Memo to Board of Works, 'Inadequate Supervision of Open Spaces', 17 June 1914, National Archives, WORKS/16/532
Design Council, *Street Scene* (London: Design Council, 1976)

'Developing the Rubberized Playground Surface', *Park Administration, Horticulture and Recreation*, 18.8 (1954), 342
Dickens, Charles, 'London Pauper Children', *Household Words*, 1850, p. 551
Dickens, Peter, 'Utopia on Trial: A Response to Alice Coleman's Comment', *International Journal of Urban and Regional Research*, 11 (1987), 118–20
'DIY Playground Equipment', *Play Times*, 15 (1979), 15
Do Something!, dir. by Ross Devenish, 1970, British Film Institute, BFI Player, https://player.bfi.org.uk/free/film/watch-do-something-1970-online [accessed 12 July 2023]
Dodds, A., 'Play and Our Young People', *Park Administration*, 28.5 (1963), 36–43
'Donation Machines Next to Playground in Regent's Park', *Camden New Journal*, 26 March 2021
'Drummond Retires', *Play Times*, 6 (1978), 3
'Dunfermline Trains Bairns in Road Sense', *Dundee Courier*, 28 June 1950, p. 2
Eagleton, G.T., 'Wanted – a Standard for Small Playgrounds', *Playing Fields Journal*, 13.2 (1953), 48–50
'Editorial', *Journal of Park Administration, Horticulture and Recreation*, 20.7 (1955), 283
'Editorial', *Park Administration, Horticulture and Recreation*, 24.5 (1959), 199
'Editorial', *Park Administration, Horticulture and Recreation*, 27.11 (1962), 21
Elgin, Lord, 'Letter to Sir Arthur Crosfield, Chairman of the NPFA', *NPFA Second Annual Report*, 1927, 12–13
Emberson, R.G., Letter to Mr Barrow, 'Regent's Park – Gloucester Green Playground', 25 September 1969, National Archives, Royal Parks, WORK 16/2299
'Evacuated Children's Holidays', *The Times*, 27 July 1940, p. 7
'Explosion in a City Schoolyard: Eight Boys Injured, Accident during Repair of Cable', *The Manchester Guardian*, 29 May 1924, p. 9
'Facilities Provided as a Result of Financial Assistance from the NPFA', *Playing Fields Journal*, 17.1 (1957), 12
'Factors in the Development of the Playing Field Movement', *Playing Fields Journal*, 19.1 (1959), 25–6
Fair Play for Children, 'Press Statement', 3 November 1978, National Archives, AT 54/159
'Fall of a School Bell: Child Killed in the Playground', *The Manchester Guardian*, 29 November 1906, p. 7
Fanning, D.M., 'Families in Flats', *British Medical Journal*, 4 (1967), 382–6
'Farleigh Hospital', *British Medical Journal*, 2 (1970), 58–9

'Fatal Accident in Charlton Kings Playground', *Gloucester Citizen*, 19 December 1912, p. 5
Firmin, Thomas, Letter to Poor Law Board, Letter, 22 August 1854, National Archives, MH 12/11000/243, fo.449–50
'For Safety's Sake: The New British Standard on Playground Equipment', *Play Times*, 13 (1979), 11
Forrest, George, 'The Giant Stride, or Flying Steps, and Its Capabilities', *Every Boy's Magazine*, 1 March 1862, pp. 122–7
'Forward to Festival of Britain', *The Architectural Review*, 110 (1951), 73–9
Friedberg, M. Paul, *Do It Yourself Playgrounds* (London: Architectural Press, 1975)
Froebel, Friedrich, *Froebel Letters: Edited with Explanatory Notes and Additional Matter by A. H. Heinemann* (Boston: Lee & Shepard, 1893)
Froebel, Friedrich, *The Education of Man*, trans. by W.N. Hailmann (Norderstedt: Vero Verlag, 2015)
Fry, Maxwell, 'The New Britain Must Be Planned', *Picture Post*, 4 January 1941, 15–18
'Future Role of the NPFA', 1970, National Archives, HLG 120/1614
Gamble, R.E., Letter to Mrs. Watson, 'Playground Safety', 7 August 1978, National Archives, AT 54/159
Garbutt, Watson, 'A Village Becomes a New Town', *Town and Country Planning*, 12.45 (1944), 22–5
Garnsey, A.H., 'Playgrounds in Europe and America', *Playing Fields Journal*, 12.3 (1952), 33–5
Garrity, E., 'Letter to Chief Commissioner of Works Regarding Indecent Offences towards Children Metropolitan Radical Federation', 22 July 1913, National Archives, WORKS/16/532
Geddes, Patrick, *Cities in Evolution: An Introduction to the Town Planning Movement and to the Study of Civics* (London: Williams & Norgate, 1915)
Gibberd, Frederick, *Harlow: The Story of a New Town* (Stevenage: Harlow Development Corporation, 1980)
Gillespie, William, 'Landscaping Our Urban Areas', *Park Administration*, 30.11 (1965), 40–43
'Girl Lost Sight of Eye from Disease Carried by Dogs', *The Times*, 25 June 1976, p. 2
Godwin, George, *Town Swamps and Social Bridges* (London: Routledge, Warnes and Routledge, 1859)
Goldfinger, Erno, 'Design for an Unidentified Playground', 1965, RIBA Collections, PA646/4(6)
Gooch, R.B., *Selection and Layout of Land for Playing Fields and Playgrounds* (London: National Playing Fields Association, 1956), National Archives, CB 4/59

Gooch, R.B., *Sketch Suggestions of Improvised Equipment for Children's Play* (London: National Playing Fields Association, 1956), London Metropolitan Archive, CLC/011/MS22287

Goodburn, J.A., Letter to Mr Poore, 'Children's Play Space – NPFA Proposals', 23 November 1972, National Archives, HLG 118/1897

Gordon, W.J., 'The London County Council and the Recreation of the People', *The Leisure Hour*, 1894, pp. 112–15

Gorvin, J.W., Letter to R.G. Emberson, 'Gloucester Green Children's Playground', 30 December 1970, National Archives, Royal Parks, WORK 16/2299

Gotch & Saunders, 'Barton Seagrave Garden Suburb Estate', 1914, Wicksteed Park Archive, uncatalogued

'Government Gymnasium', *Illustrated London News*, 29 April 1848, p. 283.

Grant, Harriet, Aamna Mohdin and Chris Michael, '"Outrageous" and "Disgusting": Segregated Playground Sparks Fury', *The Guardian*, 26 March 2019

Greater London Council, 'Toddlers' Playgrounds: Play Provision on GLC Housing Estates', 1970, London Metropolitan Archive, GLC/HG/HHM/12/S026A

Greene, A.J., 'The Open Air School Movement, 1904–1912', *The Public Health Journal*, 3.10 (1912), 547–52

Greenhalgh, Liz, and Ken Worpole, *Park Life: Urban Parks and Social Renewal* (London: Comedia/Demos, 1995)

Hall, G. Stanley, *Youth: Its Education, Regimen, and Hygiene* (New York: Appleton, 1906)

'Handwritten Note Regarding Guide to Inter-Departmental Responsibilities', 1979, National Archives, AT 60/30

Harding, Stewart, 'Towards a Renaissance in Urban Parks', *Cultural Trends*, 9.35 (1999), 1–20

Harris, G.H., 'All-Rubber Playgrounds Reduce Accident Risks', *Playing Fields*, 14.4 (1954), 36–8

Harrison, Alf T., 'A Children's Paradise: The Children's Playground', *Journal of Park Administration, Horticulture and Recreation*, 11.7 (1946), 167–71

Hart, Charles, Letter to Poor Law Commission, Letter, 14 November 1845, National Archives, MH 12/5967/153, fo.300

Hart, Ernest, 'Graveyards as Recreation Grounds', *The Times*, 20 August 1885, p. 8

Hart, M., 'Dual Use Education Playgrounds', *Parks and Recreation*, 43.8 (1978), 40–41

Harvey, Susan, 'Play in Hospital', *Mental Health*, 24.3 (1965), 121–3

Haxeltine, Michael, 'A Check List for Play Spaces', *Parks and Recreation*, 38.6 (1973), 25–9
Hayes, Rowland, Letter to Paul C. Wilson, 'Papers from New York Committee on Recreation to the New York City Mayor's Office', 29 October 1917, National Archives, RECO 1/694
Hayward, Geoffrey, Marilyn Rothenberg and Robert R. Beasley, 'Children's Play and Urban Playground Environments: A Comparison of Traditional, Contemporary, and Adventure Playground Types', *Environment and Behavior*, 6.2 (1974), 131–68
'Health and Safety Act', *Play Times*, 15 (1979), 2
Hedges, Nick, Larry Herman and Ron McCormick, 'Problems in the City' (Institute of Contemporary Arts, London, 1975)
Heiser, T.M., 'Note for File: Play Space', 18 November 1971, National Archives, AT 54/24
'Helping with Safety in Manchester', *Play Times*, 16 (1979), 2
Henderson, Robert, 'Things Made by Children', *Strand Magazine*, 1897, 752–62
Heseltine, Peter, *Regular Inspection of Children's Playgrounds* (London: Royal Society for the Prevention of Accidents, 1999)
Heseltine, Peter, *The Children's Playground: A Basic Guide* (London: Royal Society for the Prevention of Accidents, 1997)
Highway Act, 1835 http://www.legislation.gov.uk/ukpga/Will4/5-6/50/section/LXXII/enacted [accessed 6 July 2023]
Hill, Octavia, *Homes of the London Poor* (London: Macmillan, 1883)
Hill, Octavia, *Letters to Fellow-Workers 1864 to 1911*, edited by Elinor Southwood Ouvry (London: Adelphi, 1933)
Hingston, J.H., 'Play Leadership in the Borough of Ramsgate, Kent', *Park Administration, Horticulture and Recreation*, 24.9 (1960), 438–40
B. Hirst & Sons, Letter to W.J. Hepburn, 'Ne Plus Ultra Playground Equipment Catalogue to Hyde Park Superintendent', 23 September 1949, National Archives, WORKS/16/391
Hodson, D.F., Letter to R.E.K. Holmes, 'Borough of Thamesdown Response to DoE Letter on Children's Playgrounds', 4 December 1978, National Archives, AT 54/159
Hogan, Paul, *Playgrounds for Free* (Cambridge, MA: MIT Press, 1974)
Hole, Vere, *Children's Play on Housing Estates*, National Building Studies Research Paper, 39 (London: HMSO, 1966)
Hole, Vere, 'Social Effects of Planned Rehousing', *The Town Planning Review*, 30.2 (1959), 161–73
Holland, B., 'London Playgrounds', *Macmillan's Magazine* (London, 1882), XLVI edition, pp. 321–4

Holme, Anthea, and Peter Massie, *Children's Play: A Study of Needs and Opportunities* (London: Michael Joseph, 1970)
Holmes, Basil, 'More Playing Fields', *The Times*, 12 March 1920, p. 12
Holmes, Isabella M., *The London Burial Grounds: Notes on Their History from the Earliest Times to the Present Day* (London: T.F. Unwin, 1896)
Holmes, R.E.K., Letter to M. Albu, 'Safety in Children's Playgrounds', 1977, National Archives, AT 54/159
Holmes, R.E.K., Letter to Local Authority Chief Executives, 'The Need for Improved Safety in Children's Playgrounds', 31 October 1978, National Archives, AT 54/159
Home Office, 'Metropolitan Juvenile Courts Statistics for the Year 1930', 1930, National Archives, HO 45/15746
House of Commons Debate, 2 February 1965, Vol.705, Col.873, Housing Estates (Children's Play Spaces) (Hansard, 1965)
'How Austria Equips Its Children's Playgrounds', *Park Administration, Horticulture and Recreation*, 22.9 (1958), 448–9
Howard, Ebenezer, *Garden Cities of To-Morrow* (London: Swan Sonnenschein & Co, 1902)
Huddart, L.A., 'Safety on the Playground', *Playing Fields*, 13.1 (1953), 42–50
Hughes, Bob, *Evolutionary Playwork*, 2nd edn (London: Routledge, 2013)
Hughes, Bob, *Notes for Adventure Playworkers* (London: Children and Youth Action Group, 1975)
H. Hunt & Son, 'Playground Equipment – the Verdict Is Yours', *Playing Fields*, 22.3 (1962), 9
H. Hunt & Son, 'Where It All Comes From', *Park Administration*, 28.1 (1963), 1
Hunt, Jean Lee, *A Catalogue of Play Equipment* (New York: Bureau of Educational Experiments, 1918), Project Gutenberg, 28466 www.gutenberg.org/ebooks/28466 [accessed 6 July 2023]
Hyde, Amanda, '£56 for Two Hours: My Family Trip to Windsor's Extortionate New Playground', *The Telegraph*, 18 August 2023 https://www.telegraph.co.uk/travel/destinations/europe/united-kingdom/england/berkshire/windsor/family-trip-to-new-kids-playground-windsor-berkshire/ [accessed 24 November 2023]
'Hyde Park or Coney Island?', *The Times*, 7 February 1930, p. 15
'Hyde Park Playing Field: £5,000 Offer to Mr. Lansbury', *The Manchester Guardian*, 1 November 1929, p. 12
Illingworth, C., P. Brennan, A. Jay, F. Al-Rawi and M. Collick, '200 Injuries Caused by Playground Equipment', *British Medical Journal*, 4.5992 (1975), 332
'In Brief', *The Guardian*, 15 June 1940, p. 8

Ingall, C.G., 'Industrial Accident Prevention', *The Journal of State Medicine*, 35.6 (1927), 360–65
Institute of Park Administration, *Report of the 1955 Annual Conference* (London: Journal of Park Administration Ltd, 1955), Museum of English Rural Life, P2870 Box 5/39
Institute of Park Administration, *Report of the First International Congress in Public Park Administration* (London: Journal of Park Administration, 1957), RHS Lindley, 969.2 Ins
Institute of Park Administration, *Report of the Second World Congress in Public Park Administration* (London: Journal of Park Administration, 1962), RHS Lindley, 969.2 Ins
Institute of Park Administration, 'Third World Congress in Public Park and Recreation Administration Bulletin No.2', 1967, Museum of English Rural Life, SR CPRE C/1/130/2
'Insuring against Third Party Risks: Playground Accidents', *Playing Fields Journal*, 1.3 (1931), 13
'"Intensive Use" Play Park', *Playing Fields*, 37.1 (1976), 64–5
International Federation of Park Administrators, 'Bulletin', 1967, Museum of English Rural Life, SR CPRE C/1/130/2
'International Landscape Awards – Gloucester Gate Playground by LUC', *Landezine*, 2021 https://landezine-award.com/gloucester-gate-playground/ [accessed 12 July 2023]
It's Ours Whatever They Say, dir. by Jenny Barraclough, 1972, British Film Institute, BFI Player https://player.bfi.org.uk/free/film/watch-its-ours-whatever-they-say-1972-online [accessed 12 July 2023]
Jackson, Raphael, 'The Week's Good Cause: Appeal on Behalf of the National Playing Fields Association', *Radio Times*, 5 June 1936, p. 18, Radio Times Archive
Jacobs, Jane, *The Death and Life of Great American Cities: The Failure of Town Planning* (Harmondsworth: Penguin, 1964)
Jephcott, Pearl, *A Troubled Area: Notes on Notting Hill* (London: Faber and Faber, 1964)
Jephcott, Pearl, *Homes in High Flats* (Edinburgh: Oliver and Boyd, 1971)
Jesse, F. Tennyson, 'Evacuation', *The Times*, 22 September 1939, p. 6
Joint Board of Clinical Nursing Studies, 'Children's Play Panel', 1977, National Archives, DY 1/77
'Juvenile Crime', *The Times*, 13 May 1916, p. 3
Kay, James, *The Training of Pauper Children* (London: Poor Law Commissioners, 1838)
Keeling, Edward, *House of Commons Debate, 30 March 1944, Vol.398, Col.1542* (Hansard, 1944)

'Keeping up to Date with Playground Equipment', *Parks and Recreation*, 42.10 (1977), 29–40

Kellmer Pringle, Mia, Letter to J. Littlewood, 'Standards', n.d., National Archives, AT 54/24

Kenealy, Annesley, 'Playgrounds in the Parks: A Plea for the Children', *The Daily Mail*, 14 March 1907, p. 6

Kennedy, J., 'Playground Planning', *Park Administration*, 32.8 (1967), 36–8

Kinchin, Juliet, and Aidan O'Connor, *Century of the Child: Growing by Design 1900–2000* (New York: The Museum of Modern Art, 2012)

Lambert, Jack, and Jenny Pearson, *Adventure Playgrounds: A Personal Account of a Playleader's Work* (Harmondsworth: Penguin, 1974)

Lambert, Sam, 'Children Playing on the Climbing Frames in the Playground, Park Hill Estate, Sheffield', 1963, RIBA Collections, AP Box 212 Sheffield

Lambert, Sam, 'Housing, Laindon, Basildon, Essex: Open Space with Children's Play Area', 1967, RIBA Collections, AP Box 752

Land Nationalisation Society, *Report 1885–6* (London: Land Nationalisation Society, 1886)

Lansbury, George, *House of Commons Debate, 24 February 1930, Vol.235, Col.1894* (Hansard, 1930)

'Latest News – Mr Charles Melly', *John Bull*, 5 June 1858, 368

Laurie, Ian C., 'Public Parks and Spaces', in *Fifty Years of Landscape Design 1934–84*, edited by Sheila Harvey and Stephen Rettig (London: The Landscape Press, 1985), pp. 63–78

Law, Sylvia, 'Planning and the Future: A Commentary on the Debate', *The Town Planning Review*, 48.4 (1977), 365–72

Lawson, Nigel, Letter to Michael Heseltine, 'Housing Subsidy System and Project Control', 7 July 1980, National Archives, HLG 118/1897

Ledermann, Alfred, and Alfred Trachsel, *Playgrounds and Recreation Spaces*, trans. by Ernst Priefert (London: The Architectural Press, 1959)

Lee, Joseph, 'Play for Home', *The Playground*, 6.5 (1912), 146–58

Leeds Civic Society, 'House and Town Planning Exhibition Programme', 1918, Wicksteed Park Archive, PRG-3004

Letchworth Dramatic Society, 'A Variety Entertainment', 1914, Garden City Collection, LBM4007.18

Levy, Geoffrey, 'Playground Monsters', *Daily Mail*, 16 May 1990, p. 6

Lipton, Marcus, *House of Commons Debate, 13 March 1953 Vol.512 Col.1735* (Hansard, 1953)

Littlewood, J., Letter to Mr. Oddy, 'Play Space Standards', 27 January 1978, National Archives, AT 54/159

Littlewood, J., Letter to T.M. Heiser, 'Play Space', 12 November 1971, National Archives, AT 54/24
'Liverpool's Adventure Playgrounds', *Park Administration, Horticulture and Recreation*, 24.10 (1960), 515–17
Local Government Board for Scotland, *Report of the Women's House-Planning Committee* (Edinburgh: HMSO, 1918)
London and Greater London Playing Fields Association, 'Children's Playgrounds Exhibition and Conference Press Release', 1954, London Metropolitan Archives, LCC/CL/PK/01/038
London County Council, 'Ceremony of Opening Little Dorrit's Playground, Southwark', 1902, London Metropolitan Archives, LCC/CL/CER/03
London County Council, 'County of London Parks, Open Spaces and Commons', 1892, London Metropolitan Archives, LCC/CL/PK/01/104
London County Council, 'Experimental Play Equipment: Sketch Layouts', 1959, London Metropolitan Archives, GLC/HG/HHM/12/S026A
London County Council, 'Parks and Open Spaces, Descriptions, By-Laws, Acts of Parliament, Regulations', 1894, London Metropolitan Archives, LCC/CL/PK/01/104
London County Council, 'Parks and Open Spaces: Regulations Relating to Games, Together with Particulars of the Facilities Afforded for General Recreation', 1915, LSE Library, 421 (129D)
London County Council, 'Play Parks', 1963, London Metropolitan Archives, LCC/PUB/11/01/119
London County Council, 'Playgrounds – Dry', 1927, London Metropolitan Archives, LCC/CL/PK/01/038
London County Council, 'Playgrounds – Equipment', 1962, London Metropolitan Archives, GLC/HG/HHM/12/S026A
London County Council, 'Recreational Facilities', 1935, London Metropolitan Archives, LCC/PK/GEN/02/003
London County Council, 'Regulations Relating to the Playing of Games at Parks and Open Spaces under the Control of the Council', 1904, LSE Library, 421 (129A)
London County Council, 'Report as to the Condition of Victoria Park', 1893, London Metropolitan Archives, LCC/CL/PK/01/104
London County Council, 'Report of the Parks and Open Spaces Committee, 16 May', 1893, London Metropolitan Archives, LCC/CL/PK/01/104
London County Council, 'Return of the Names and Wages of and Work Performed by All Persons Employed in the Council's Parks and Also of the Respective Areas Devoted to Gardens, Lawns, Fields and

Playgrounds and the Extent of Conservatories in Such Parks', 1889, London Metropolitan Archives, LCC/CL/PK/01/104
London County Council, 'Spa Fields Extension', 1951, London Metropolitan Archives, GLC/RA/D2G/04/091
London County Council, 'Unsupervised Play Space on Housing Estates', 1959, London Metropolitan Archives, GLC/HG/HHM/12/S026A
'London County Council', *The Times*, 6 February 1889, p. 11
'London County Council', *The Times*, 30 March 1898, p. 15
London County Council Education Department. Letter to Bailiff of the Royal Parks, 'Child Molestation Cases', 27 January 1932, National Archives, WORKS/16/532
'London Playing Fields Society', *The Times*, 3 June 1899, p. 14
London Standing Conference of Housing Estate Community Groups, 'Newsletter', 1960, London Metropolitan Archives, GLC/HG/HHM/12/S026A
Lord Chelmsford Opens First Children's Playground to Be Erected under Miners' Welfare Fund in Colliery Districts, 1926, British Pathé Archive, 634.18 https://www.britishpathe.com/video/lord-chelmsford-1 [accessed 6 July 2023]
'Lord Meath's Memories', *The Times Literary Supplement*, 10 May 1923, p. 317
Loudon, J.C., *An Encyclopædia of Gardening* (London: Longman, Rees, Orme, Brown, Green and Longman, 1835)
Lovatt, J., A.N. Worden, J. Pickup and C.E. Brett, 'The Fattening of Pigs on Swill Alone: A Municipal Enterprise', *Empire Journal of Experimental Agriculture*, 11 (1943), 182–90
Lowndes, G., 'The Cost of Traffic Accidents, the Case for Playing Fields from a New Angle', *Playing Fields Journal*, 4.1 (1936), 19–22
Luke, Lord, 'Festival of Britain 1951 NPFA to Play Important Part', *Playing Fields Journal*, 9.4 (1949), 202
Lynch, Kevin, *The Image of the City* (Cambridge, MA: MIT Press, 1960)
Macaulay, Rose, *Pleasure of Ruins* (London: Weidenfeld & Nicolson, 1953)
Macey, John P., 'Problems of Flat Life', *Official Architecture and Planning*, 22.1 (1959), 35–8
Mackenzie, W. Leslie, *Scottish Mothers and Children* (Dunfermline: Carnegie United Kingdom Trust, 1917)
Maclaren, Archibald, *A Military System of Gymnastic Exercises* (London: HMSO, 1868)
Madge, Charles, 'Planning for People', *The Town Planning Review*, 21.2 (1950), 131–44
Maizels, Joan, *Two to Five in High Flats* (London: Housing Centre, 1961)

Major, Joshua, *The Theory and Practice of Landscape Gardening* (London: Longman, Brown, Green and Longmans, 1852)
Markwell, Susan, and Neil Ravenscroft, *The Public Provision of Children's Playgrounds* (Reading: University of Reading, 1996)
Marriott, W.T., *The Necessity of Open Spaces and Public Playgrounds in Large Towns* (Manchester: Manchester and Salford Sanitary Association, 1862), Wellcome Collection
Marron, J.A., and C.L. Senn, 'Dog Feces: A Public Health and Environmental Problem', *Journal of Environmental Health*, 37 (1974), 239–43
Masterman, Charles, 'Realities at Home', in *The Heart of Empire: Discussions of Problems of Modern City Life in England, with an Essay on Imperialism*, edited by Charles Masterman (London: Fisher Unwin, 1902), pp. 1–52
Maud, P., 'Recreation in Public Parks and Open Spaces', *Playing Fields Journal*, 1.2 (1930), 7–13
Mayhew, Henry, *London Labour and the London Poor* (London, 1851), I
Mays, John Barron, *Adventure in Play: The Story of the Rathbone Street Adventure Playground* (Liverpool: Liverpool Council of Social Service, 1957)
McGovern, Bernard, *Playleadership* (London: Faber and Faber, 1973)
McHattie, John W., *Report on Public Parks, Gardens and Open Spaces* (Edinburgh: City of Edinburgh, 1914), RHS Lindley, 999 4C EDI
McKendrick, John, 'Playgrounds in the Built Environment', *Built Environment*, 25.1 (1999), 5–10
McKendrick, John, Anna Fielder and Michael Bradford, 'Privatization of Collective Play Spaces in the UK', *Built Environment*, 25.1 (1999), 44–57
McMillan, Margaret, *The Nursery School* (London: J.M. Dent & Sons, 1921)
McMillan, R.C., 'Changes in the Playground', *The Times*, 12 July 1957, p. 11
McNab, Archie, 'Equipping Children's Playgrounds', *Design*, 159 (1962), 64–8
Mearns, Andrew, *The Bitter Cry of Outcast London: An Inquiry into the Condition of the Abject Poor (1883)* (London: Frank Cass, 1970)
Measuring Worth, 'Five Ways to Compute the Relative Value of a U.K. Pound Amount, 1270 to Present', 2023, www.measuringworth.com/ukcompare [accessed 6 June 2023]
Meath, Earl of, *House of Lords Debate, 9 June 1890, Vol.345, Col.264–267* (Hansard, 1890)
Meath, Earl of, *Public Parks of America*, Report to the Parks and Open Space Committee, London County Council, 1890, London Metropolitan Archives, LCC/PUB/02/01/066

Meath, Earl of, 'Public Playgrounds for Children', *The Nineteenth Century*, 34 (1893), 267–71

Mero, Everett B., *American Playgrounds: Their Construction, Equipment, Maintenance and Utility* (Boston: School of Education, Harvard University, 1908)

Metropolitan Boroughs' Standing Joint Committee, 'Proposed Closing of Streets to Traffic for Use of Children', 1934, National Archives, MEPO 2/7803

Metropolitan Free Drinking Fountain Association, *Annual Report*, 1865, London Metropolitan Archives, ACC/3168/017

Metropolitan Police, 'Memorandum from "M" Division, Tower Bridge Station Regarding Juvenile Courts, 4 August', 1932, National Archives, HO 45/15746

Metropolitan Public Garden, Boulevard, and Playground Association, *Second Annual Report* (London, 1884), US National Library of Medicine, archive.org, 101200449

Metropolitan Public Gardens Association, *Annual Report*, 1932, London Metropolitan Archive, CLC/011/MS22290

Metropolitan Public Gardens Association, *Eighteenth Annual Report* (London, 1900)

Metropolitan Public Gardens Association, *Twenty Second Annual Report*, 1904, The Guildhall Library, ST 317

Milton Keynes Development Corporation, 'Play in a New City', *Playing Fields*, 34.1 (1973), 26–31

'Miners' Welfare: A Model Playground in Wales', *The Times*, 2 June 1930, p. 11

Ministry of Education, *The Youth Service in England and Wales* (London: HMSO, 1960)

Ministry of Health, *Housing Manual* (London: HMSO, 1949)

Ministry of Health, 'Housing Manual', 1944, National Archives, HLG 110/10

Ministry of Housing and Local Government, *Families Living at High Density: A Study of Estates in Leeds, Liverpool and London*, Design Bulletin, 21 (London: HMSO, 1970)

Ministry of Housing and Local Government, *Homes for Today and Tomorrow* (London: HMSO, 1961)

Ministry of Housing and Local Government, *Houses 1952: Second Supplement to the Housing Manual 1949* (London: HMSO, 1952)

Ministry of Housing and Local Government, *Houses 1953*, Circular 54/53 (London: HMSO, 1953)

Ministry of Housing and Local Government, *Houses 1953: Third Supplement to the Housing Manual 1949* (London: HMSO, 1953)

Ministry of Housing and Local Government, 'Housing Handbook', 1957, National Archives, HLG 31/11

Ministry of Housing and Local Government, *Housing Standards, Costs and Subsidies*, Circular 36/67 (London: HMSO, 1967)

Ministry of Reconstruction Advisory Council, *Women's Housing Sub-Committee Final Report* (London: HMSO, 1919), National Archives, RECO 1/629

Minoprio, Anthony, 'Crawley New Town', *Town and Country Planning*, 16.64 (1949), 215–21

Mitchell, Eleanor, 'Planning for Children's Play', *Town and Country Planning*, 34.8–9 (1966), 418–21

Mitchell, Mary, 'Birmingham Parks', *Park Administration*, 28.5 (1963), 47

Mitchell, Mary, 'Birmingham Playgrounds', *Playing Fields*, 21.4 (1961), 40–41

Mitchell, Mary, 'Birmingham Playgrounds', *Playing Fields*, 24.2 (1964), 29–30

Mitchell, Mary, 'Landscaping of Housing Areas', *Official Architecture and Planning*, 25.4 (1962), 193–96

Mitton, Roger, and Elizabeth Morrison, *A Community Project in Notting Dale* (London: Allen Lane, 1972)

Model Traffic Area No. 1 (Pathetones, 1938), British Pathé Archive, PT440 www.britishpathe.com/video/model-traffic-area-no-1 [accessed 6 July 2023]

Montgomery, M.W., 'The Week's Good Cause: Appeal on Behalf of the Glasgow and District Playing-Fields Association', *Radio Times*, 24 May 1929, p. 403, Radio Times Archive

Moore, Robin C., 'Playgrounds at the Crossroads', in *Public Places and Spaces*, edited by Irwin Altman and Ervin H. Zube (Boston, MA: Springer, 1989), pp. 83–120

Moore, Robin C., and Donald Young, 'Childhood Outdoors: Toward a Social Ecology of the Landscape', in *Children and the Environment*, edited by Irwin Altman and Joachim F. Wohlwill, Human Behaviour and Environment: Advances in Theory and Research (New York: Plenum Press, 1978), III, 83–128

'More Playgrounds, Please, Say Parents', *Playing Fields Journal*, 12.1 (1952), 38

'Mothers Want Safer Streets', *Daily Mail*, 12 October 1937, p. 11

'Mr. Chas. Wicksteed's Generosity: Kettering Club's Appreciation, Mr Wicksteed Silences His Critics', *The Kettering Guardian*, 15 July 1921, p. 6

'Mr Lansbury and the London Parks', *The Manchester Guardian*, 11 February 1930, p. 6

'Mr. Lansbury and the Parks: Hyde Park or Coney Island? Reply to Critics', *The Observer*, 9 February 1930, p. 17
'Mr Lansbury and the Parks: Peace or Playground: The Serpentine Pavilion', *The Times*, 12 February 1930, p. 15
'Mr Lansbury and the Parks: The New Technique: Correspondents' Protests', *The Times*, 8 February 1930, p. 13
'Mr. Lansbury and the Royal Parks: Criticism and Defence in Commons', *The Manchester Guardian*, 25 February 1930, p. 5
Mulcaster, Richard, *Positions* (London: Longmans, Green & Co., 1888)
Murphy, Thomas, 'The Evolution of Amusement Machines', *Journal of the Royal Society of Arts*, 99.4855 (1951), 791–806
National Children's Playground Association, 'The Five Million Club, Minutes of Executive Committee, 17 October', 1950, National Archives, CB 1/76
National Playing Fields Association, 'Adventure Playgrounds at Farleigh Hospital', 1971, National Archives, CB 1/63
National Playing Fields Association, 'Annual Report and Accounts', 1976, Museum of English Rural Life, SR CPRE C/1/73/2
National Playing Fields Association, *Best Play: What Play Provision Should Do for Children* (London: National Playing Fields Association, 2000)
National Playing Fields Association, 'DoE Circular 79/72 Children's Play Space: Note by the Children's Play Officer', 1972, National Archives, CB 1/61
National Playing Fields Association, 'Draft Constitution', n.d., London Metropolitan Archive, LCC/CL/PK/01/042
National Playing Fields Association, *First Annual Report* (National Playing Fields Association, 1927), National Archives, CB 4/1
National Playing Fields Association, 'Institute of Playleadership Minutes, 9 February', 1970, National Archives, CB 1/64
National Playing Fields Association, 'King George's Fields County Register 1931–1965', 1965, National Archives, CB 2/24
National Playing Fields Association, 'Memorandum on the Recent Work of the Association', 1942, Museum of English Rural Life, SR CPRE C/1/73/1
National Playing Fields Association, 'Minutes of Play Leadership Committee 23 April', 1936, National Archives, CB 1/54
National Playing Fields Association, 'Minutes of the Ad Hoc Committee to Enquire into the Provision of Play Space on New Housing Estates, 17 May', 1960, National Archives, CB 1/64
National Playing Fields Association, 'Minutes of the Children's Playground Committee on 17 March', 1954, National Archives, CB 1/59

National Playing Fields Association, 'Minutes of the Children's Playground Committee on 18 November', 1954, National Archives, CB 1/59
National Playing Fields Association, 'Minutes of the First Meeting of the Children's Playground Technical Subcommittee on 4 November', 1952, National Archives, CB 1/68
National Playing Fields Association, 'Minutes of the Sub-Committee on Play Leadership 14 December', 1928, National Archives, CB 1/54
National Playing Fields Association, 'Play Leadership in the United States', 1933, National Archives, CB 1/54
National Playing Fields Association, 'Playground and Play Leadership Committee Minutes', 1963, National Archives, CB 1/55
National Playing Fields Association, *Playgrounds for Blocks of Flats* (London: National Playing Fields Association, 1953), Museum of English Rural Life, P2870 Box 5/12
National Playing Fields Association, *Playgrounds for Blocks of Flats*, 6th edn (London: National Playing Fields Association, 1974), National Archives, CB 4/76
National Playing Fields Association, 'Register of Grants Given under the Physical Training and Recreation Act 1937', 1942, National Archives, CB 1/83
National Playing Fields Association, 'Report of Adventure Playground Conference', 1956, National Archives, CB 1/67
National Playing Fields Association, 'Report of Conference on Play Leadership', 1933, National Archives, CB 1/54
National Playing Fields Association, 'Report of Proceedings at Inaugural Meeting of the National Playing Fields Association', 1925, National Archives, CB 1/1
National Playing Fields Association, *Second Annual Report* (National Playing Fields Association, 1928), National Archives, CB 4/1
National Playing Fields Association, *Sixth Annual Report* (National Playing Fields Association, 1932), National Archives, CB 4/1
National Playing Fields Association, 'Steering Committee on Adventure Playgrounds', 1953, National Archives, CB 1/53
National Playing Fields Association, *Survey of Urban Playing Facilities* (London: National Playing Fields Association, 1951)
National Playing Fields Association, *Third Annual Report* (National Playing Fields Association, 1929), National Archives, CB 4/1
National Playing Fields Association, *Towards a Safer Adventure Playground* (London: National Playing Fields Association, 1980)
National Union of Women's Suffrage Societies, 'Letchworth and District Society', 1912, Garden City Collection, LBM2988

'Need for Village Playing Fields', *Playing Fields Journal*, 2.7 (1934), 312
Neill, A.S., *Summerhill: A Radical Approach to Child-Rearing* (Harmondsworth: Penguin, 1968)
'New Flats in Liverpool', *The Times*, 21 June 1935, p. 13
'New Playfield Ideas from All over the World: Fascinating Facts in Report on Children's Playground Exhibition', *Playing Fields Journal*, 15.1 (1955), 53
Newson, John, and Elizabeth Newson, *Four Years Old in an Urban Community* (London: Allen & Unwin, 1968)
'Newspaper Chat', *Examiner*, 12 June 1825, p. 745
Nicholson, Simon, 'How NOT to Cheat Children: The Theory of Loose Parts', *Landscape Architecture*, 62.1 (1971), 30–34
Nicholson, Simon, 'The Theory of Loose Parts, an Important Principle for Design Methodology', *Studies in Design Education Craft & Technology*, 4.2 (1972), 5–14
Nicol, Ian, Letter to Mr Moss, 'New Subsidy Arrangements: Playspace Provision', 6 October 1980, National Archives, HLG 118/1897
'Ninth Exhibition of Park Equipment, Machinery and Materials', *Park Administration, Horticulture and Recreation*, 26.11 (1961), 46–8
Niven, David, 'The Parks and Open Spaces of London', in *London of the Future*, edited by Aston Webb (London: The London Society, 1921), pp. 235–50
'Northumberland Road Adventure Playground and the Health and Safety at Work Act, 1974', *Play Times*, 14 (1979), 2
'Official Inspection of the Manchester Public Parks', *Manchester Guardian*, 19 August 1846, p. 6
O'Kelly, J.D., 'Parks and Playgrounds', *Landscape and Garden*, 5 (1938), 94–5
Olmsted, Frederick Law, 'Boston: Charleston Playground: General Plan', 1891, Artstor/University of California, San Diego
Olmsted, Frederick Law, *Public Parks* (Brookline, MA, 1902) https://archive.org/details/publicparksbeinooolmsgoog/page/n7/mode/2up [accessed 20 February 2024]
O'Malley, Jan, *The Politics of Community Action: A Decade of Struggle in Notting Hill* (Nottingham: Spokesman Books, 1977)
'Open Spaces in Hackney', *Daily News*, 18 April 1890, p. 5
'Open Spaces in Parliament', *The Times*, 23 February 1885, p. 4
'Opening of Burbury Street Recreation Ground', *Birmingham Daily Post*, 3 December 1877, p. 8
'Opening of Clissold Park', *The Standard*, 25 July 1889, p. 3

'Opening of Stamford Park, Altrincham', *Manchester Times*, 30 October 1880, p. 7
'Opening of the Portsmouth Public Park', *Hampshire Telegraph*, 29 May 1878, p. 3
Opie, Iona, and Peter Opie, *Children's Games in Street and Playground* (Oxford: Clarendon Press, 1969)
Ordnance Survey, 'Manchester and Salford Town Plan Sheet 21', 1850
'Out of the Way: Our Gang', *New Society*, 8 February 1973, pp. 305–6
P.A. Management Consultants, 'Organisation and Control', 1967, Wicksteed Park Archive, uncatalogued
'Paddington's Municipal Piggery: Two Years of Remarkable Progress', *Journal of Park Administration, Horticulture and Recreation*, 8.3 (1943), 35
Paris, G.E., 'A Children's Playground and Model Traffic Area', *Playing Fields Journal*, 5.2 (1939), 89–96
'Park Safety Still Causing Concern', *Evening Telegraph*, 26 February 1968, Wicksteed Park Archive
Paton Watson, J., and Patrick Abercrombie, *A Plan for Plymouth* (Plymouth: Underhill, 1943)
Pepler, George L., 'Open Spaces', *The Town Planning Review*, 10.1 (1923), 11–24
Pepper, Simon, and Peter Richmond, 'Upward or Outward? Politics, Planning and Council Flats, 1919–1939', *The Journal of Architecture*, 13.1 (2008), 53–90
'Peril of Playground Death Traps', *The Guardian*, 12 June 1978, p. 3
Pertwee, W.R., 'Designing Children's Gymnasia', *Landscape and Garden*, 6 (1939), 28–31
'Pets Danger to Children Playing in Parks', *The Times*, 23 November 1973, p. 5
Pettigrew, W.W., *Handbook of Manchester Parks and Recreation Grounds* (Manchester: Manchester City Council, 1929)
Pettigrew, W.W., *Municipal Parks: Layout, Management and Administration* (London: Journal of Park Administration, 1937), RHS Lindley, 969.2 PET
Pettigrew, W.W., 'Park Superintendent's Report Book', trans. by Anne Bell and Andrew Bell, 1908, Cardiff Council Parks Service, uncatalogued
Pettigrew, W.W., 'The Manchester and Salford Gardens Guild', *Radio Times*, 14 October 1927, p. 78, Radio Times Archive
Pettigrew, W.W., 'The Northern Garden', *Radio Times*, 24 April 1931, p. 243, Radio Times Archive

'Photo of Wicksteed's Munition Girls F.C. Team', n.d., Wicksteed Park Archive, uncatalogued

'Physical Recreation, Manchester and Playing Fields', *The Manchester Guardian*, 7 November 1913, p. 16

'Planning', *The Architect and Building News*, 1957, 477–81

'Planning Children's Playgrounds', *The Times*, 12 June 1954, p. 3

'Planning Standards Criticised', *The Guardian*, 17 March 1966, p. 2

'Play Chute Negligence Case', *Playing Fields*, 17.1 (1957), 48–9

Play England, *Developing an Adventure Playground: The Essential Elements*, Practice Briefing (London: Play England, 2009)

'Play Fields for Youths: New Manchester Movement', *Manchester Courier*, 9 July 1907, p. 9

'Play Leadership Film', *Playing Fields Journal*, 16.4 (1956), 25

Play Safety Forum, *Managing Risk in Play Provision: A Position Statement* (London: Children's Play Council, 2002)

'Play Space Standards: List of Equipment Taken from Wicksteed Catalogue', 1971, National Archives, AT 54/24

'Playground Equipment (Charles Wicksteed)', 1927, Cape Town Archives Repository, 3/CT 4/1/4/71 B410/4

'Playground and General Recreation Society', *Daily News*, 3 June 1858, p. 3

'Playground Mishaps Injure 20,000', *The Guardian*, 6 May 1976, p. 6

'Playground Probe Reveals Dangers in Swings and Slides', *The Sunday Times*, 4 June 1972, p. 3

'Playground and Recreation Society', *Lloyd's Illustrated*, 22 May 1859, p. 7

'Playground Risks', *Playing Fields Journal*, 3.1 (1934), 3–4

'Playgrounds in Birmingham', *The Architect and Building News*, 218.24 (1960), 767–8

'Playing Fields for the People', *The Manchester Guardian*, 4 April 1925, p. 7

'Playing-Grounds for Young People', *The Times*, 4 April 1925, p. 17

Playstyle Ltd, 'The Complete Playground: Playcubes', *Playing Fields*, 34.1 (1973), 10

'Politics and Play: Some Comments by a Play Worker', *Play Times*, 6 (1978), 2

Potrony, Albert, *Equal Play*, 2021, BALTIC

Potts, L., Letter, 'Children's Playgrounds', 21 May 1965, National Archives, Royal Parks, WORK 16/2299

Prentice Mawson, Edward, 'Public Parks and Playgrounds: A New Conception', *Parks, Golf Courses and Sports Grounds*, 1 (1935), 7–8

Prior, W.H., 'The Queen in Victoria Park', *Illustrated London News*, 12 April 1873, p. 349

'Prof. Voelker Pentonville Gymnasium', *The Lady's Magazine*, 31 July 1827, 392

'Publications Available from the NPFA', *Playing Fields*, 37.4 (1976), 49–50
Pybus, John, *House of Commons Debate, Royal Parks and Pleasure Gardens, 24 February 1930 Vol.235 Col.1909* (Hansard, 1930)
'Quarry Hill Flats, Playground, Kitson House', 1939, Leeds Central Library, D LIE Quarry (12) https://www.leodis.net/viewimage/98774 [accessed 6 July 2023]
'Questions in the Commons: The Royal Parks: Mr. Lansbury's Plans Challenged', *The Manchester Guardian*, 18 February 1930, p. 6
Rainwater, Clarence E., *The Play Movement in the United States: A Study of Community Recreation* (Chicago: University of Chicago Press, 1922)
Raphael, David, 'Grounds for Playful Renaissance', *Landscape Architecture*, 65.3 (1975), 329–30
Rapoport, Amos, 'The Home Range of the Child', *Ekistics*, 45.272 (1978), 378
'Ratepayers Petition Council for Children's Playground', *Western Times*, 24 April 1936, p. 7
Reaney, Mabel Jane, 'A Director of Play', *The Manchester Guardian*, 11 March 1919, p. 4
Reaney, Mabel Jane, *The Place of Play in Education* (London: Methuen & Co., 1927)
Reaney, Mabel Jane, 'The Urgent Need for Trained Play Leaders, a Paper given at the Conference on Play Leadership at the Institute of Education, London, 21 July', 1933, National Archives, CB 1/54
'Recreation for Evacuees', *The Guardian*, 18 December 1939, p. 10
Recreation Grounds Act, 1859 https://www.legislation.gov.uk/ukpga/Vict/22/27/contents/enacted [accessed 6 July 2023]
'Reductions in Capital Expenditure from Public Funds', *Playing Fields*, 25.4 (1965), 26
Reeves, Will R., 'Report of Committee on Street Play', *The Journal of Educational Sociology*, 4.10 (1931), 607–18
Richardson, Benjamin Ward, *Hygeia: A City of Health* (London: Macmillan, 1876)
Richardson, J., 'A Century of Playing Fields Progress', *Playing Fields*, 6.4 (1946), 155–60
Richardson, J., 'The Provision of Open Spaces in Slum Clearance Areas and Congested Districts', *Journal of Park Administration, Horticulture and Recreation*, 4.5 (1939), 125–9
'Road Safety for Children', *The Manchester Guardian*, 1 May 1956, p. 14
Roth, Mathias, *Gymnastic Exercises without Apparatus According to Ling's System*, 7th edn (London: A.N. Myers & Co., 1887)
Routledge, D.A., R. Repetto-Wright and C. I. Howarth, 'The Exposure of Young Children to Accident Risk as Pedestrians', *Ergonomics*, 17.4 (1974), 457–80

'Royal 12-Day Tour Begins', *The Times*, 28 June 1958, p. 4
Royal Parks, 'Children's Playgrounds: Gifts of Equipment Offers and Acceptances', 1923, National Archives, WORKS/16/1705
Royal Parks, 'General Layout Plan for Gloucester Green Children's Playground', 1969, National Archives, WORK 16/2299
Royal Parks, 'Kensington Gardens Children's Playground', 1909, National Archives, WORKS/16/391
Royal Parks, 'Memoranda in Response to an Article in the Times, Dated 18 November 1933, on a Recreation Ground Swing Accident in Walthamstow', 1933, National Archives, WORKS/16/846
Royal Parks, 'Park Keepers' Reports of Offences against Children', 1932, National Archives, WORKS/16/532
Royal Parks, 'Prevention of Offences against Children (Various Minutes and Notes)', 1913, National Archives, WORKS/16/532
Royal Parks, 'Primrose Hill Children's Playground, Gymnasium and Lavatories', 1938, National Archives, WORKS/16/1670
Royal Parks, 'St. James's Park. Children's Playground and Paddling Facilities', 1930, National Archives, WORKS/16/1504
'Runaway Horse Kills Kid', *The Manchester Guardian*, 4 December 1924, p. 9
Saarinen, Aline B., 'Playground: Function and Art', *New York Times*, 4 July 1954, p. 4
'Safety on Play Areas', *Play Times*, 1 (1977), 11
'Salford Town Council Park', *Manchester Examiner and Times*, 13 November 1850, p. 6
Satterthwaite, Bob, Letter to R.E.K. Holmes, 'The Need for Improved Safety in Children's Playgrounds', 24 November 1978, National Archives, AT 54/159
Saunders, Peter Bruce, 'Oral History Interview, Reference to Playing in Damaged Buildings at 00:08:00', 1999, Imperial War Museum, Sound 18748
Sayers, Philip, 'Whence and Whither? From Parks Superintendents to Leisure Planners', *Park Administration*, 31.12 (1966), 18–19
Seeley, Ivor, *Outdoor Recreation and the Urban Environment* (London: Palgrave Macmillan, 1973)
'Seen at the Public Health Exhibition', *Journal of Park Administration, Horticulture and Recreation*, 1 (1936), 247
Sexby, John James, *The Municipal Parks, Gardens, and Open Spaces of London: Their History and Associations* (London: Elliot Stock, 1898)
Shackell, Aileen, Nicola Butler, Phil Doyle and David Ball, *Design for Play: A Guide to Creating Successful Play Spaces* (London: Department

for Children, Schools and Families and Department for Culture, Media and Sport, 2008)

Sharp, Evelyn, *The London Child* (London: John Lane, 1927)

Sharp, Evelyn, *The Ministry of Housing and Local Government* (London: Allen & Unwin, 1969)

'Shelter Play-Centres', *The Guardian*, 18 October 1941, p. 8

Sheppard, F.H.W., 'The Crown Estate in Kensington Palace Gardens: Individual Buildings', in *Survey of London*, Northern Kensington (London: London County Council, 1973), XXXVII, 162–93 http://www.british-history.ac.uk/survey-london/vol37/pp162-193 [accessed 16 November 2023]

Shier, Harry, *Adventure Playgrounds: An Introduction* (London: National Playing Fields Association, 1984)

Silver, Henry, 'Ragged Playgrounds', *Punch*, 1 May 1858, 181

'Six New Pieces of Playground Equipment', *Evening Telegraph*, 29 July 1964, p. 8

'Slum Clearance at Poplar', *The Times*, 30 August 1934, p. 6

'Slum Clearance in London', *The Times*, 8 March 1938, p. 21

Smith, Peter, 'Time to Give Play a New Priority', *Municipal Review*, 44.519 (1973), 78–81

SMP Landscapes, 'Play Area Layout, Scheme B, Wicksteed Park Kettering', 1979, Wicksteed Park Archive, uncatalogued

Sørensen, Carl Theodore, *Parkpolitik i Sogn Og Købstad* (Copenhagen: Christian Ejlers Forlag, 1931)

Southwark, Bishop of, *House of Lords Debate, 14 February 1928, Vol. 70, Col. 92* (Hansard, 1928)

'Special Report: From Five to Fourteen', *Play Times*, 6 (1978), 8–9

'Speeches from the Meeting Held by the National Playing Fields Association', *Radio Times*, 3 July 1925, p. 57, Radio Times Archive

Spencer, Heath and George Ltd, Letter to D. Campbell, 'Sketch for Regent's Park Superintendent', 1 May 1924, National Archives, WORKS/16/1705

Spencer, Heath and George Ltd, 'Playground Apparatus, 1932 Improved Models', *Playing Fields Journal*, 1.8 (1932), xiii

Spencer, Heath and George Ltd, 'Playground Catalogue', 1927, National Archives, WORKS/16/1705

Spencer, Heath and George Ltd, 'Sketches of Regulation Swing Frame, Patent Safety Giant Stride & See-Saw', n.d., National Archives, WORKS/16/1705

Spicker, Paul, 'Poverty and Depressed Estates: A Critique of Utopia on Trial', *Housing Studies*, 2.4 (1987), 283–92

Stanley, Mr., *House of Commons Debate, 19 January 1981, Vol. 997, Col. 66, Play Space* (Hansard, 1981)
'Sting in the Tail: Man's Best Friend Can Also Be One of Children's Worst Enemies', *The Guardian*, 8 February 1989, p. 27
'Street Accidents', *Playing Fields Journal*, 1.8 (1932), 12
'Street Accidents to Children', *Playing Fields Journal*, 1.3 (1931), 19–20
'Street and Driving Accidents: The Coroner and Children's Playgrounds', *The Manchester Guardian*, 18 June 1907, p. 14
'Success of "Safety First" Campaign', *The Times*, 29 March 1924, p. 15
Sudell, Richard, 'How Can We Make Our Parks Brighter?', *Municipal Journal and Public Works Engineer*, 46.2300 (1937), 397–8
Sudell, Richard, 'Park Design for Modern Needs', *Journal of Park Administration, Horticulture and Recreation*, 14.10 (1950), 343–6
Sudell, Richard, 'Wanted – More Play Leaders', *Playing Fields Journal*, 12.1 (1952), 20–24
Sutcliffe Engineering, 'Seat for a Swing', 1978, Espacenet Patent Search, GB1535728A
'Swedish-Inspired Play Equipment Experiment', *Evening Telegraph*, 29 July 1964, p. 6, Wicksteed Park Archive, uncatalogued
Taylor, Ian, and Paul Walton, 'Hey, Mister, This Is What We Really Do ...', in *Vandalism*, edited by Colin Ward (London: Architectural Press, 1973), pp. 91–5
'Teaching Children Road Safety', *The Times*, 28 July 1938, p. 11
Terrill, Simon, and Assemble, *The Brutalist Playground*, 2015, RIBA
'The Advantage of Taking a Short Cut through a Court', *Punch*, 4 June 1859, 233
The Battle for Powis Square, 1975, London Community Video Archive, CVA0019b https://lcva.gold.ac.uk/videos/5e5522d7c899b32870ce69d2 [accessed 12 July 2023]
'The Childless City', *The Observer*, 10 September 1939, p. 11
'The Children of St Helena', *Machinery Lloyd*, 29 (1957), 1–2
'The Cry of the Children', *Playing Fields Journal*, 2.1 (1932), 4–5
'The Gardens of Lincoln's Inn Fields', *The Times*, 22 May 1860, p. 12
'The Growing Variety in Playground Furniture', *The Municipal and Public Services Journal*, 75 (1967), 1272–3
'The Gymnasium, Primrose Hill', *The Penny Illustrated Paper*, 2 May 1863, p. 4
'The Hathern Playground', *Leicester Chronicle*, 14 January 1837, p. 4
'The Heedless Child: What He Means to the Motorist', *The Manchester Guardian*, 9 October 1928, p. 8
'The Job of the Games Warden', *Playing Fields Journal*, 12.1 (1952), 25–6

The Leisure of the People: A Handbook, Being the Report of the National Conference Held at Manchester, November 17th–20th, 1919 (Manchester: Conference Committee, 1919)
'The Library Shelf', *Official Architecture and Planning*, 23.3 (1960), 135
'The London Gymnastic Institution', *The Times*, 5 January 1827, p. 4
'The Need for Play', *Hospital*, 66 (1919), 52
'The Ocean Wave', *The Times*, 3 January 1890, p. 4
'The Playing Fields Association', *The Manchester Guardian*, 25 July 1925, p. 7
'The Playing Fields Society', *The Manchester Guardian*, 19 October 1908, p. 10
'The Regent's Park Gymnasium', *The Standard*, 21 August 1878, p. 2
The Scottish Government, *Play Strategy for Scotland: Our Vision* (Edinburgh: Scottish Government, 2013)
'The Showman World', *The Era*, 6 May 1899, p. 18
'The Swing Accident in an Edinburgh Playground', *Edinburgh Evening News*, 26 January 1905, p. 4
'The Trifler', *The Sunday Times*, 11 April 1880, p. 7
'The Use of Closed Streets as Playgrounds', *Playing Fields Journal*, 3.2 (1935), 35–6
'The X-Certificate Playgrounds', *Play Times*, 2 (1977), 6–7
Thompson, Paul, 'The War with Adults', *Oral History*, 3.2 (1975), 29–38
Thornton Rutter, H., 'The Chronicle of the Car', *Illustrated London News*, 6 August 1938, p. 256
'Tottenham's Contribution to Road Safety', *Journal of Park Administration, Horticulture and Recreation*, 3.2 (1938), 95–6
'Traffic and Children: Expecting Too Much Wisdom', *The Manchester Guardian*, 2 October 1936, p. 8
'Training While They Play: Tottenham's Model Traffic Area', *Journal of Park Administration, Horticulture and Recreation*, 12.3 (1947), 85–8
'Triodetic Aluminium Playdome', *Park Administration*, 31.12 (1966), 39
Turner, Tom, and Simon Rendel, *London Landscape Guide* (Dartford: Landscape Institute, 1983)
Ulrich, W.C., Letter to P.R.O.s, 'NPFA Advice on Children's Playgrounds', 1968, National Archives, AT 54/24
University Hall Settlement, 'Memorandum and Articles of Association' (London, 1895), LSE Library, FOLIO FHV/G60
'Urban Areas Need More Games Facilities', *Playing Fields Journal*, 11.2 (1951), 32–5
Valangin, Francis de, *A Treatise on Diet, or the Management of Human Life* (London: Pearch, 1768)

'Vandals Leave £1.8bn Bill and a Trail of Fear', *The Sunday Times*, 13 August 1989, p. 5
'Various Cuttings from Trade Journals and Newspapers', 1966, Wicksteed Park Archive, uncatalogued
Vile, T.H., 'The Week's Good Cause: Appeal on Behalf of the Monmouthshire Branch of the National Playing Fields Association', *Radio Times*, 8 July 1938, p. 25, Radio Times Archive
'Voelker's Gymnastics', *Examiner*, 11 December 1825, p. 787
Voyce, Tom, 'The Week's Good Cause: Appeal on Behalf of the Gloucestershire Playing Fields Association', *Radio Times*, 12 February 1937, p. 22, Radio Times Archive
Walker, Donald, *British Manly Exercises* (London: Hurst, 1834)
Ward, Colin, 'Adventure Playground', *Freedom*, 6 September 1958, pp. 3–4
Ward, Colin, 'Adventure Playground: A Parable of Anarchy', *Anarchy*, 7 (1961), 193–201
Ward, Colin, *Anarchy in Action* (London: Allen & Unwin, 1973)
Ward, Colin, *The Child in the City*, 2nd edn (London: Bedford Square Press, 1990)
Ward, Colin, *The Child in the Country* (London: Bedford Square Press, 1990)
'Welcome to Your First Issue ...', *Play Times*, 1 (1977), 3
Welsh Assembly Government, *Play Policy* (Cardiff: Welsh Assembly, 2002)
Wesley, Reginald, 'Play Leadership in the City of Belfast', *Park Administration, Horticulture and Recreation*, 24.8 (1960), 378–80
Westergaard, John, and Ruth Glass, 'A Profile of Lansbury', *The Town Planning Review*, 25.1 (1954), 33–58
White, Christopher, 'Rampage of the Tiny Vandals, Aged 10', *Daily Mail*, 30 April 1985, p. 1
'Why Do We Let Dogs Foul Our Streets?', *British Medical Journal*, 1 (1976), 1486
Wicksteed, Charles, *A Plea for Children's Recreation after School Hours and after School Age* (Kettering: Wicksteed Charitable Trust, 1928)
Wicksteed, Charles, *Bygone Days and Now: A Plea for Co-Operation between Labour, Brains and Capital* (London: Williams & Northgate, 1929)
Wicksteed, Charles, 'Concrete Cottages', *The Machine Tool Review*, 1920, Wicksteed Park Archive, uncatalogued
Wicksteed, Charles, 'National Coal: The Farce of Nationalisation Exposed.', n.d., Wicksteed Park Archive, MGA-3006
Wicksteed, Charles, *Our Mother Earth: A Short Statement of the Case for Land Nationalisation* (London: Swan Sonnenschein & Co, 1892)

Wicksteed, Charles, *The Land for the People: How to Obtain It and How to Manage It* (London: William Reeves, 1885)

Wicksteed, Charles, 'The Pity of It: Thoughtless Picnic Parties in the Wicksteed Park', *The Kettering Leader*, 29 July 1921, p. 5

Wicksteed, Hilda M., *Charles Wicksteed* (London: J. M. Dent & Sons, 1933)

Wicksteed, Joseph H., *The Challenge of Childhood: An Essay on Nature and Education* (London: Chapman & Hall, 1936)

'Wicksteed: Our Works Are Now Fully Occupied on 100% War Work', *Journal of Park Administration, Horticulture and Recreation*, 5.11 (1941), front cover

'Wicksteed Park: Kettering Clubmen's Appreciation of the Founder', *The Kettering Leader*, 15 July 1921, p. 7

Wicksteed Village Trust, 'An Account of the Wicksteed Park and Trust', 1936, Wicksteed Park Archive, BRC-1906

Wicksteed Village Trust, 'Annual Accounts 1916–48', Wicksteed Park Archive, uncatalogued

Wicksteed Village Trust, 'Hi! There's a Big Smile for Everyone to Wear at Wicksteed Park', n.d., Wicksteed Park Archive, BRC-1821

Wicksteed Village Trust, Letter to J. Brandon-Jones, Letter, 4 October 1946, Wicksteed Park Archive, LET-1044

Wicksteed Village Trust, 'Minute Book, 1920–1935', Wicksteed Park Archive, uncatalogued

Wicksteed Village Trust, 'The Wicksteed Park, Kettering', n.d., Wicksteed Park Archive, BRC-1199

Wicksteed Village Trust, 'The Wicksteed Park Souvenir', n.d., London Metropolitan Archive, CLC/011/MS22290

Wicksteed Village Trust, 'Wicksteed Park; 140 Acres of Fun and Freedom', 1970, Wicksteed Park Archive, BRC-1793

Wicksteed Village Trust, 'Wicksteed Park Nature Trail', 1971, Wicksteed Park Archive, PRM-1883

Wilderspin, Samuel, *A System of Education for the Young* (London: Hodson, 1840)

Wilkinson, Tom, 'Duchess's Vision Sees World's Biggest Play Park Opened', *Evening Standard*, 24 May 2023 https://www.standard.co.uk/news/uk/duchess-northumberland-b1083434.html [accessed 24 November 2023]

Williams-Ellis, Clough, 'Biased Opinions', *Journal of Park Administration, Horticulture and Recreation*, 12.1 (1947), 15–17

Willis, Margaret, 'High Blocks of Flats: A Social Survey', 1955, London Metropolitan Archives, GLC/HG/HHM/12/S026A Toddlers' Playgrounds

Willis, Margaret, 'Sandpits: A Social Survey', 1951, University of Edinburgh, PJM/LCC/D/6
Willis, Margaret, 'Toddlers' Playgrounds: An Enquiry into the Reactions of Mothers to the Experimental Play Equipment for Pre-School Aged Children', 1959, London Metropolitan Archive, GLC/HG/HHM/12/S026A
Winship, A.K., 'Editorial – Playing in Sand', *The Journal of Education*, 70.16 (1909), 436
Wodehouse, P.G., *Love among the Chickens* (New York: Circle, 1909)
Wood, Denis, 'Free the Children! Down with Playgrounds!', *McGill Journal of Education*, 12.2 (1977), 227–42
Wood, Walter, *Children's Play and Its Place in Education* (London: Kegan Paul, Trench, Trübner & Co., 1913)
Woodruff, A.W., 'Toxocariasis as a Public Health Problem', *Environmental Health*, 84 (1976), 29–31
Wroth, Warwick, *The London Pleasure Gardens of the Eighteenth Century* (London: Macmillan, 1896)
Wuellner, Lance H., 'Forty Guidelines for Playground Design', *Journal of Leisure Research*, 11.1 (1979), 4–14
Young, Arthur, *A Tour in Ireland, 1776–1779*, 2nd edn (London: Cassell, 1897)
Young, Michael, 'Kinship and Family in East London', *Man*, 54.210 (1954), 137–9
'Young Offenders in War Time', *The Times*, 13 March 1916, p. 5

Secondary Sources

Andersson, Sven-Ingvar, and Steen Høyer, *C. Th. Sørensen, Landscape Modernist*, trans. by Anne Whiston Spirn (Copenhagen: Danish Architectural Press, 2001)
Andrews, Howard F., 'Home Range and Urban Knowledge of School-Age Children', *Environment and Behavior*, 5.1 (1973), 73–86
Antler, Joyce, 'Mitchell, Lucy Sprague (1878–1967), Educator', *American National Biography* (Oxford: Oxford University Press, 1999)
Armytage, W.H.G., *Heavens below: Utopian Experiments in England 1560–1960* (London: Routledge and Kegan Paul, 1961)
Atkinson, Harriet, *The Festival of Britain: A Land and Its People* (London: Tauris, 2012)
Auerbach, Jeffrey A., *The Great Exhibition of 1851: A Nation on Display* (New Haven: Yale University Press, 1999)
Austin, Linda M., 'Children of Childhood: Nostalgia and the Romantic Legacy', *Studies in Romanticism*, 42.1 (2003), 75–98

Baigent, Elizabeth, 'Chubb, Sir Lawrence Wensley (1873–1948), Environmental Campaigner', *Oxford Dictionary of National Biography* (Oxford: Oxford University Press, 2004)

Bailey, Richard, *A.S. Neill* (London: Bloomsbury, 2013)

Baldwin, David, 'Major, Joshua (1786–1866), Landscape Gardener and Designer', *Oxford Dictionary of National Biography* (Oxford: Oxford University Press, 2004)

Ball, David, Tim Gill and Bernard Spiegal, *Managing Risk in Play Provision: Implementation Guide* (London: National Children's Bureau, 2012)

Balmforth, Nick, 'Drummond Abernethy OBE (1913–88)', *Journal of Playwork Practice*, 1 (2014), 101–3

Bartrip, P.W.J., 'Hart, Ernest Abraham (1835–1898), Medical Journalist', *Oxford Dictionary of National Biography* (Oxford: Oxford University Press, 2004)

Beckett, Ian, 'The Nation in Arms, 1914–1918', in *A Nation in Arms: A Social Study of the British Army in the First World War*, edited by Ian Beckett and Keith Simpson (Barnsley: Pen and Sword, 2004), pp. 1–36

Beckett, Ian, Timothy Bowman and Mark Connelly, *The British Army and the First World War* (Cambridge: Cambridge University Press, 2017)

Beevers, Robert, *The Garden City Utopia: A Critical Biography of Ebenezer Howard* (Basingstoke: Macmillan, 1988)

Black, Lawrence, and Hugh Pemberton, 'The Benighted Decade? Reassessing the 1970s', in *Reassessing 1970s Britain*, edited by Lawrence Black, Hugh Pemberton and Pat Thane (Manchester: Manchester University Press, 2013)

Bluemel, Kristin, and Michael McCluskey, eds., *Rural Modernity in Britain: A Critical Intervention* (Edinburgh: Edinburgh University Press, 2018)

Borsay, Peter, *A History of Leisure: The British Experience since 1500* (Basingstoke: Palgrave Macmillan, 2006)

Borsay, Peter, 'Pleasure Gardens and Urban Culture in the Long Eighteenth Century', in *The Pleasure Garden, from Vauxhall to Coney Island*, edited by Jonathan Conlin (Philadelphia: University of Pennsylvania Press, 2013), pp. 49–77

Bosselmann, Peter, 'Landscape Architecture as Art: C. Th. Sørensen. A Humanist', *Landscape Journal*, 17.1 (1998), 62–9

Boughton, John, *Municipal Dreams: The Rise and Fall of Council Housing* (London: Verso, 2019)

Bradley, Kate, 'Creating Local Elites: The University Settlement Movement, National Elites and Citizenship in London, 1884–1940', in

In Control of the City: Local Elites and the Dynamics of Urban Politics, 1800–1960, edited by Stefan Couperus, Christianne Smit and Dirk Jan Wolffram, Groningen Studies in Cultural Change, 28 (Leuven: Peeters, 2007), pp. 81–92

Brehony, Kevin, 'A "Socially Civilising Influence"? Play and the Urban "Degenerate"', *Paedagogica Historica*, 39 (2003), 87–106

Brehony, Kevin, 'Transforming Theories of Childhood and Early Childhood Education: Child Study and the Empirical Assault on Froebelian Rationalism', *Paedagogica Historica*, 45 (2009), 585–604

Brück, Joanna, 'Landscapes of Desire: Parks, Colonialism, and Identity in Victorian and Edwardian Ireland', *International Journal of Historical Archaeology*, 17 (2013), 196–223

Burkhalter, Gabriela, ed., *The Playground Project* (Zurich: JRP|Ringier, 2016)

Carroll, P., O. Calder-Dawe, K. Witten and L. Asiasiga, 'A Prefigurative Politics of Play in Public Places: Children Claim Their Democratic Right to the City through Play', *Space and Culture*, 22.3 (2019), 294–307

Cavallo, Dominick, *Muscles and Morals: Organized Playgrounds and Urban Reform, 1880–1920* (Philadelphia: University of Pennsylvania Press, 1981)

Colton, Ruth, 'Savage Instincts, Civilising Spaces: The Child, the Empire and the Public Park, c.1880–1914', in *Children, Childhood and Youth in the British World*, edited by Shirleene Robinson and Simon Sleight (Basingstoke: Palgrave, 2016), pp. 255–70

Conekin, Becky, *'The Autobiography of a Nation': The 1951 Festival of Britain* (Manchester: Manchester University Press, 2003)

de Coninck-Smith, Ning, *Natural Play in Natural Surroundings: Urban Childhood and Playground Planning in Denmark, c.1930–1950*, Working Papers in Child and Youth Culture, 6 (Odense: University of Southern Denmark, 1999)

de Coninck-Smith, 'Where Should Children Play? City Planning Seen From Knee-Height: Copenhagen 1870 to 1920', *Children's Environments Quarterly*, 7.4 (1990), 54–61

Conlin, Jonathan, 'Vauxhall on the Boulevard: Pleasure Gardens in London and Paris, 1764–1784', *Urban History*, 35 (2008), 24–47

Conlin, Jonathan, 'Vauxhall Revisited: The Afterlife of a London Pleasure Garden, 1770–1859', *Journal of British Studies*, 45 (2006), 718–43

Conway, Hazel, 'Everyday Landscapes: Public Parks from 1930 to 2000', *Garden History*, 28 (2000), 117–34

Conway, Hazel, *People's Parks: The Design and Development of Victorian Parks in Britain* (Cambridge: Cambridge University Press, 1991)

Cook, Lynn, 'The 1944 Education Act and Outdoor Education: From Policy to Practice', *History of Education*, 28.2 (1999), 157–72

Cowman, Krista, '"A Peculiarly English Institution": Work, Rest, and Play in the Labour Church', *Studies in Church History*, 37 (2002), 357–67

Cowman, Krista, 'Open Spaces Didn't Pay Rates: Appropriating Urban Space for Children in England after WW2', in *Städtische Öffentliche Räume: Planungen, Aneignungen, Aufstände 1945–2015 (Urban Public Spaces: Planning, Appropriation, Rebellions 1945–2015)*, edited by Christoph Bernhardt (Stuttgart: Franz Steiner Verlag, 2016), pp. 119–40

Cowman, Krista, 'Play Streets: Women, Children and the Problem of Urban Traffic, 1930–1970', *Social History*, 42 (2017), 233–56

Cowman, Krista, '"The Atmosphere Is Permissive and Free": The Gendering of Activism in the British Adventure Playgrounds Movement, ca. 1948–70', *Journal of Social History*, 53.1 (2019), 218–41

Crawford, Elizabeth, 'Wilkinson, Fanny Rollo (1855–1951), Landscape Gardener', *Oxford Dictionary of National Biography* (Oxford: Oxford University Press, 2008)

Cross, Gary, and John K. Walton, *The Playful Crowd: Pleasure Places in the Twentieth Century* (New York: Columbia University Press, 2005)

Csepely-Knorr, Luca, and Mária Klagyivik, 'From Social Spaces to Training Fields: Evolution of Design Theory of the Children's Public Sphere in Hungary in the First Half of the Twentieth Century', *Childhood in the Past*, 13.2 (2020), 93–108

Csepely-Knorr, Luca, and Amber Roberts, 'Towards a "Total Environment" for Children: Michael Brown's Landscapes for Play', in *Landscape and Children* (presented at the FOLAR Annual Symposium, Museum of English Rural Life, 2019)

Cunningham, Hugh, *Children and Childhood in Western Society since 1500* (London: Routledge, 2005)

Darling, Elizabeth, '"The Star in the Profession She Invented for Herself": A Brief Biography of Elizabeth Denby, Housing Consultant', *Planning Perspectives*, 20.3 (2005), 271–300

Darling, Elizabeth, 'What the Tenants Think of Kensal House: Experts' Assumptions versus Inhabitants' Realities in the Modern Home', *Journal of Architectural Education*, 53.3 (2000), 167–77

Daunton, M.J., 'How to Pay for the War: State, Society and Taxation in Britain, 1917–24', *The English Historical Review*, 111 (1996), 882–919

Davies, Rhodri, *Public Good by Private Means* (London: Alliance Publishing Trust, 2015)

Davin, Anna, *Growing up Poor* (London: Rivers Oram Press, 1996)
Dawson, Graham, *Soldier Heroes: British Adventure, Empire, and the Imagining of Masculinities* (London: Routledge, 1994)
Deloria, Philip J., *Playing Indian* (New Haven: Yale University Press, 1998)
Dickinson, Elizabeth, 'The Misdiagnosis: Rethinking "Nature-Deficit Disorder"', *Environmental Communication*, 7.3 (2013), 315–35
Donaldson, Peter, *Sport, War and the British* (London: Routledge, 2020)
Downs, Annabel, 'Sudell, Richard (1892–1968), Landscape Architect and Author', *Oxford Dictionary of National Biography* (Oxford: Oxford University Press, 2009)
Dreher, Nan, 'The Virtuous and Verminous: Turn of the Century Moral Panics in London's Public Parks', *Albion*, 29 (1997), 246–67
Driver, Felix, 'Moral Geographies: Social Science and the Urban Environment in Mid-Nineteenth Century England', *Transactions of the Institute of British Geographers*, 13 (1988), 275–87
Dyhouse, Carol, 'The British Federation of University Women and the Status of Women in Universities, 1907–1939', *Women's History Review*, 4.4 (1995), 465–85
Elyot, Thomas, *The Book Named the Governor* (London: Dent, 1965)
English, Jim, 'Empire Day in Britain, 1904–1958', *The Historical Journal*, 49 (2006), 247–76
Fishman, Robert, *Urban Utopias in the Twentieth Century: Ebenezer Howard, Frank Lloyd Wright, Le Corbusier* (Cambridge, MA: MIT Press, 1982)
Fletcher, Robert, 'Connection with Nature Is an Oxymoron: A Political Ecology of "Nature-Deficit Disorder"', *The Journal of Environmental Education*, 48.4 (2017), 226–33
Floud, Roderick, *An Economic History of the English Garden* (London: Allen Lane, 2019)
Francis, Martin, 'The Domestication of the Male? Recent Research on Nineteenth- and Twentieth-Century British Masculinity', *The Historical Journal*, 45 (2002), 637–52
Fraser, Derek, *The Evolution of the British Welfare State* (London: Palgrave Macmillan, 2017)
Gagen, Elizabeth, 'An Example to Us All: Child Development and Identity Construction in Early 20th-Century Playgrounds', *Environment and Planning A: Economy and Space*, 32.4 (2000), 599–616
Gagen, Elizabeth, 'Landscapes of Childhood and Youth', in *A Companion to Cultural Geography*, edited by James Duncan, Nuala Johnson, and Richard Schein (Oxford: Blackwell, 2004), pp. 404–19

Gagen, Elizabeth, 'Playing the Part: Performing Gender in America's Playgrounds', in *Children's Geographies: Playing, Living, Learning*, edited by Sarah Holloway and Gill Valentine (London: Routledge, 2000), pp. 213–29

Gagen, Elizabeth, 'Too Good to Be True: Representing Children's Agency in the Archives of Playground Reform', *Historical Geography*, 29 (2001), 53–64

Gärtner, Niko, 'Administering "Operation Pied Piper": How the London County Council Prepared for the Evacuation of Its Schoolchildren 1938–1939', *Journal of Educational Administration and History*, 42.1 (2010), 17–32

Gill, Tim, *No Fear: Growing Up in a Risk Averse Society* (London: Calouste Gulbenkian Foundation, 2007)

Gill, Tim, *Urban Playground: How Child-Friendly Planning and Design Can Save Cities* (London: RIBA, 2021)

Glasheen, Lucie, 'Bombsites, Adventure Playgrounds and the Reconstruction of London: Playing with Urban Space in Hue and Cry', *The London Journal*, 44.1 (2019), 54–74

Graham, Philip, *Susan Isaacs: A Life Freeing the Minds of Children* (London: Routledge, 2009)

Greenhalgh, James, *Reconstructing Modernity: Space, Power, and Governance in Mid-Twentieth Century British Cities* (Manchester: Manchester University Press, 2018)

Greenhalgh, James, 'The New Urban Social History? Recent Theses on Urban Development and Governance in Post-War Britain', *Urban History*, 47.3 (2020), 535–45

Grosvenor, Ian, and Kevin Myers, '"Dirt and the Child": A Textual and Visual Exploration of Children's Physical Engagement with the Urban and the Natural World', *History of Education*, 49.4 (2020), 517–35

Grundlingh, Louis, 'Municipal Modernity: The Politics of Leisure and Johannesburg's Swimming Baths, 1920s to 1930s', *Urban History*, 49.4 (2022), 771–90

Gunn, Simon, 'The Rise and Fall of British Urban Modernism', *Journal of British Studies*, 49.4 (2010), 849–69

Gunn, Simon, and Alastair Owens, 'Nature, Technology and the Modern City: An Introduction', *Cultural Geographies*, 13 (2006), 491–6

Hall, Alison, 'The Shelter Photographs 1968–1972: Nick Hedges, the Representation of the Homeless Child and a Photographic Archive' (unpublished thesis, University of Birmingham, 2016)

Hannikainen, Matti O., *The Greening of London 1920–2000* (Farnham: Ashgate, 2016)

Harrington, Ralph, 'On the Tracks of Trauma: Railway Spine Reconsidered', *Social History of Medicine*, 16.2 (2003), 209–23
Harwood, Elain, 'Post-War Landscape and Public Housing', *Garden History*, 28.1 (2000), 102–16
Harwood, Elain, 'Review: The New Brutalist Image 1949–55 and The Brutalist Playground', *Journal of the Society of Architectural Historians*, 75.1 (2016), 117–19
Hendrick, Harry, *Child Welfare, England 1872–1989* (London: Routledge, 1994)
Hendrick, Harry, 'Constructions and Reconstructions of British Childhood: An Interpretative Survey, 1800 to the Present', in *Constructing and Reconstructing Childhood*, edited by Allison James and Alan Prout (London: Routledge, 2015), pp. 29–53
Hickman, Clare, 'Care in the Countryside: The Theory and Practice of Therapeutic Landscapes in the Early Twentieth Century', in *Gardens and Green Spaces in the West Midlands since 1700*, edited by Malcolm Dick and Elaine Mitchell (Hatfield: University of Hertfordshire Press, 2018), pp. 160–85
Hickman, Clare, 'To Brighten the Aspect of Our Streets and Increase the Health and Enjoyment of Our City', *Landscape and Urban Planning*, 118 (2013), 112–19
Highmore, Ben, 'Playgrounds and Bombsites: Postwar Britain's Ruined Landscapes', *Cultural Politics*, 9 (2013), 323–36
Hines, Michael, '"They Do Not Know How To Play": Reformers' Expectations and Children's Realities on the First Progressive Playgrounds of Chicago', *The Journal of the History of Childhood and Youth*, 10 (2017), 206–27
Historic England, *Water Chute at Wicksteed Park, National Heritage List for England, 1437706* (Swindon: Historic England, 2016)
Hogarth, Margaret, 'Campbell, Dame Janet Mary (1877–1954), Medical Officer', *Oxford Dictionary of National Biography* (Oxford: Oxford University Press, 2006)
Hollow, Matthew, 'Suburban Ideals on England's Interwar Council Estates', *Garden History*, 39.2 (2011), 203–17
Hollow, Matthew, 'Utopian Urges: Visions for Reconstruction in Britain, 1940–1950', *Planning Perspectives*, 27.4 (2012), 569–85
Holman, Bob, *Champions for Children: The Lives of Modern Child Care Pioneers* (Bristol: Policy Press, 2001)
Holt, Richard, *Sport and the British: A Modern History* (Oxford: Clarendon Press, 1992)
Howell, Ocean, 'Play Pays: Urban Land Politics and Playgrounds in the United States, 1900–1930', *Journal of Urban History*, 34 (2008), 961–94

Huggins, Mike, and Jack Williams, *Sport and the English 1918–1939* (London: Routledge, 2006)
Hulme, Tom, *After the Shock City: Urban Culture and the Making of Modern Citizenship* (Woodbridge: Boydell Press, 2019)
Hulme, Tom, 'Putting the City Back into Citizenship: Civics Education and Local Government in Britain, 1918–45', *Twentieth Century British History*, 26.1 (2015), 26–51
Jessen, Asbjørn, and Anne Tietjen, 'Assembling Welfare Landscapes of Social Housing: Lessons from Denmark', *Landscape Research*, 46.4 (2021) https://doi.org/10.1080/01426397.2020.1808954 [accessed 9 Feb. 2024]
Johnston, Scott, 'Courting Public Favour: The Boy Scout Movement and the Accident of Internationalism, 1907–29', *Historical Research*, 88.241 (2015), 508–29
Jones, Claire L., *The Medical Trade Catalogue in Britain, 1870–1914* (London: University of Pittsburgh Press, 2013)
Jones, Karen R., 'Green Lungs and Green Liberty: The Modern City Park and Public Health in an Urban Metabolic Landscape', *Social History of Medicine*, 35.4 (2022), 1200–1222
Jones, Karen R., '"The Lungs of the City": Green Space, Public Health and Bodily Metaphor in the Landscape of Urban Park History', *Environment and History*, 24 (2018), 39–58
Jones, Peter, 'Re-Thinking Corruption in Post-1950 Urban Britain: The Poulson Affair, 1972–1976', *Urban History*, 39.3 (2012), 510–28
Jones, Stephen, *Workers at Play: A Social and Economic History of Leisure 1918–39* (London: Routledge, 1986)
Jordan, Harriet, 'Mawson, Thomas Hayton (1861–1933), Landscape Architect', *Oxford Dictionary of National Biography* (Oxford: Oxford University Press, 2010)
Jouhki, Essi, 'Politics in Play: The Playground Movement as a Socio-Political Issue in Early Twentieth-Century Finland', *Paedagogica Historica* (2023) https://doi.org/10.1080/00309230.2022.21554811–21 [accessed 9 Feb. 2024]
Kane, Josephine, *The Architecture of Pleasure: British Amusement Parks 1900–1939* (Farnham: Ashgate, 2013)
Kotlaja, M., E. Wright and A. Fagan, 'Neighborhood Parks and Playgrounds: Risky or Protective Contexts for Youth Substance Use?', *Journal of Drug Issues*, 48.4 (2018), 657–75
Kozlovsky, Roy, 'Adventure Playgrounds and Postwar Reconstruction', in *Designing Modern Childhoods: History, Space, and the Material Culture of Children; An International Reader*, edited by Marta Gutman

and Ning de Coninck-Smith (New Jersey: Rutgers University Press, 2007), pp. 171–90

Kozlovsky, Roy, *The Architectures of Childhood: Children, Modern Architecture and Reconstruction in Postwar England* (Farnham: Ashgate, 2013)

Laakkonen, Simo, 'Asphalt Kids and the Matrix City: Reminiscences of Children's Urban Environmental History', *Urban History*, 38 (2011), 301–23

Lambert, David, 'Rituals of Transgression in Public Parks in Britain, 1846 to the Present', in *Performance and Appropriation: Profane Rituals in Gardens and Landscapes*, edited by Michel Conan (Dumbarton Oaks: Harvard University Press, 2007), pp. 195–210

Lambert, David, *The Park Keeper* (London: English Heritage, 2005)

Lang, Ruth, 'The Sociologist within: Margaret Willis and the London County Council Architect's Department' (presented at the Architecture and Bureaucracy: Entangled Sites of Knowledge Production and Exchange, Brussels, 2019)

Langhamer, Claire, *Women's Leisure in England 1920–60* (Manchester: Manchester University Press, 2000)

Larrivee, Shaina D., 'Playscapes: Isamu Noguchi's Designs for Play', *Public Art Dialogue*, 1.1 (2011), 53–80

Lasdun, Susan, *The English Park: Royal, Private and Public* (London: Andre Deutsch, 1991)

Laybourn, Keith, and David Taylor, 'Traffic Accidents and Road Safety: The Education of the Pedestrian and the Child, 1900–1970', in *The Battle for the Roads of Britain: Police, Motorists and the Law, c.1890s to 1970s*, edited by Keith Laybourn and David Taylor (London: Palgrave Macmillan, 2015), pp. 149–85

Layton-Jones, Katy, 'Manufactured Landscapes: Victorian Public Parks and the Industrial Imagination', in *Gardens and Green Spaces in the West Midlands since 1700*, edited by Malcolm Dick and Elaine Mitchell (Hatfield: University of Hertfordshire Press, 2018), pp. 120–37

Layton-Jones, Katy, *National Review of Research Priorities for Urban Parks, Designed Landscapes, and Open Spaces: Final Report*, Research Report Series, 4 (London: English Heritage, 2014)

Layton-Jones, Katy, and Robert Lee, *Places of Health and Amusement: Liverpool's Historic Parks and Gardens* (Swindon: English Heritage, 2008)

Lefaivre, Liane, and Ingeborg de Roode, eds., *Aldo van Eyck: The Playgrounds and the City* (Amsterdam: Stedelijk Museum, 2002)

Lucas, John, 'A Centennial Retrospective: The 1889 Boston Conference on Physical Training', *Journal of Physical Education, Recreation & Dance*, 60.9 (1989), 30–33

Lynch, Gordon, 'Pathways to the 1946 Curtis Report and the Post-War Reconstruction of Children's Out-of-Home Care', *Contemporary British History*, 34 (2020), 22–43

MacAlister, Ian, and John Elliott, 'Gotch, John Alfred (1852–1942)', *Oxford Dictionary of National Biography* (Oxford: Oxford University Press, 2004)

Malchow, H.L., 'Public Gardens and Social Action in Late Victorian London', *Victorian Studies*, 29 (1985), 97–124

Mangan, J.A., *'Manufactured' Masculinity: Making Imperial Manliness, Morality and Militarism* (London: Routledge, 2012)

Marmaras, Emmanuel, and Anthony Sutcliffe, 'Planning for Post-war London: The Three Independent Plans, 1942–3', *Planning Perspectives*, 9.4 (1994), 431–53

Martin, Michael, Andrea Jelić, and Tenna Doktor Olsen Tvedebrink, 'Children's Opportunities for Play in the Built Environment: A Scoping Review', *Children's Geographies*, 21.6 (2023), 1154–70 https://doi.org/10.1080/14733285.2023.2214505 [accessed 9 Feb. 2024]

Matless, David, *Landscape and Englishness* (London: Reaktion Books, 1998)

Matthew, H.C.G., 'Masterman, Charles Frederick Gurney (1873–1927), Politician and Author', *Oxford Dictionary of National Biography* (Oxford: Oxford University Press, 2015)

Mazumdar, Pauline, 'Burt, Sir Cyril (1883–1971)', *Oxford Dictionary of National Biography* (Oxford: Oxford University Press, 2004)

McQueeney, Kevin G., 'More than Recreation: Black Parks and Playgrounds in Jim Crow New Orleans', *Louisiana History*, 60.4 (2019), 437–78

Mellor, Leo, *Reading the Ruins: Modernism, Bombsites and British Culture* (Cambridge: Cambridge University Press, 2011)

Mergen, Bernard, 'Children and Nature in History', *Environmental History*, 8 (2003), 643–69

Moggridge, Hal, 'Allen, Marjory, Lady Allen of Hurtwood (1897–1976), Landscape Architect and Promoter of Child Welfare', *Oxford Dictionary of National Biography* (Oxford: Oxford University Press, 2007)

Moran, Joe, 'Crossing the Road in Britain, 1931–1976', *The Historical Journal*, 49 (2006), 477–96

Morgan, W. John, 'The Miners' Welfare Fund in Britain 1920–1952', *Social Policy and Administration*, 24 (1990), 199–211

Moshenska, Gabriel, 'Children in Ruins', in *Ruin Memories: Materiality, Aesthetics and the Archaeology of the Recent Past*, edited by Bjørnar Olsen and Þóra Pétursdóttir (Abingdon: Routledge, 2014), pp. 230–49
Murnaghan, Ann Marie, 'Exploring Race and Nation in Playground Propaganda in Early Twentieth Century Toronto', *International Journal of Play*, 2 (2013), 134–46
Murnaghan, Ann Marie, and Laura Shillington, 'Digging Outside the Sandbox: Ecological Politics of Sand and Urban Children', in *Children, Nature, Cities*, edited by Ann Marie Murnaghan and Laura Shillington (London: Routledge, 2016)
Natural England, *The Children's People and Nature Survey for England* (London: Office for National Statistics, 2022)
Norman, Nils, *An Architecture of Play: A Survey of London's Adventure Playgrounds* (London: Four Corners Books, 2003)
O'Reilly, Carole, 'From "The People" to "The Citizen": The Emergence of the Edwardian Municipal Park in Manchester, 1902–1912', *Urban History*, 40 (2013), 136–55
Orme, Nicholas, *Medieval Children* (New Haven: Yale University Press, 2001)
Orrock, Amy, 'Homo Ludens: Pieter Bruegel's Children's Games and the Humanist Educators', *Journal of Historians of Netherlandish Art*, 4.2 (2012), 1–21
Ortolano, Guy, *Thatcher's Progress: From Social Democracy to Market Liberalism through an English New Town* (Cambridge: Cambridge University Press, 2019)
Osborn, F.J., 'Pepler, Sir George Lionel (1882–1959)', *Oxford Dictionary of National Biography* (Oxford: Oxford University Press, 2004)
Pearson, Geoffrey, '"A Jekyll in the Classroom, a Hyde in the Street": Queen Victoria's Hooligans', in *Crime and the City*, edited by David Downes (London: Macmillan, 1989), pp. 10–35
Pemberton, Neil, 'The Burnley Dog War: The Politics of Dog-Walking and the Battle over Public Parks in Post-Industrial Britain', *Twentieth Century British History*, 28.2 (2017), 239–67
Pfister, Gertrud, 'Cultural Confrontations: German Turnen, Swedish Gymnastics and English Sport – European Diversity in Physical Activities from a Historical Perspective', *Culture, Sport, Society*, 6 (2003), 61–91
Philips, Deborah, *Fairground Attractions: A Genealogy of the Pleasure Ground* (London: Bloomsbury, 2012)
Pitsikali, A., and R. Parnell, 'Fences of Childhood: Challenging the Meaning of Playground Boundaries in Design', *Frontiers of Architectural Research*, 9.3 (2020), 656–69

'Playground, n.', *Oxford English Dictionary* (Oxford: Oxford University Press, 2006)
Pomfret, David, 'The City of Evil and the Great Outdoors: The Modern Health Movement and the Urban Young, 1918–40', *Urban History*, 28 (2001), 405–27
Pussard, Helen, 'Historicising the Spaces of Leisure: Open-Air Swimming and the Lido Movement in England', *World Leisure Journal*, 49.4 (2007), 178–88
Ravetz, Alison, *Model Estate: Planned Housing at Quarry Hill, Leeds* (London: Croom Helm, 1974)
Read, Jane, 'Gutter to Garden: Historical Discourses of Risk in Interventions in Working Class Children's Street Play', *Children and Society*, 25 (2011), 421–34
Reeder, David A., 'The Social Construction of Green Space in London Prior to the Second World War', in *The European City and Green Space: London, Stockholm, Helsinki and St. Petersburg, 1850–2000*, edited by Peter Clark (Aldershot: Ashgate, 2006), pp. 41–67
Richards, Thomas, *The Commodity Culture of Victorian England: Advertising and Spectacle, 1851–1914* (London: Verso, 1991)
Riskin, Jessica, 'Machines in the Garden', *Republics of Letters*, 1.2 (2010), 16–43
Roberts, Judith, 'The Gardens of Dunroamin: History and Cultural Values with Specific Reference to the Gardens of the Inter-war Semi', *International Journal of Heritage Studies*, 1.4 (1996), 229–37
Robinson, Lucy, *Gay Men and the Left in Post-War Britain* (Manchester: Manchester University Press, 2007)
Rosenzweig, Roy, and Elizabeth Blackmar, *The Park and the People: A History of Central Park* (Ithaca: Cornell University Press, 1992)
Ruff, Allan, *The Biography of Philips Park, Manchester 1846–1996*, School of Planning and Landscape Occasional Paper 56 (Manchester: University of Manchester Press, 2000)
Rutherford, Vanessa, 'Muscles and Morals: Children's Playground Culture in Ireland, 1836–1918', in *Leisure and the Irish in the Nineteenth Century*, edited by Leeann Lane and William Murphy (Liverpool: Liverpool University Press, 2016), pp. 61–79
Saumarez Smith, Otto, *Boom Cities: Architect-Planners and the Politics of Radical Urban Renewal in 1960s Britain* (Oxford: Oxford University Press, 2019)
Saumarez Smith, Otto, 'The Inner City Crisis and the End of Urban Modernism in 1970s Britain', *Twentieth Century British History*, 27 (2016), 578–98

Saumarez Smith, Otto, 'The Lost World of the British Leisure Centre', *History Workshop Journal*, 88 (2019), 180–203

Savage, Mike, *Identities and Social Change in Britain since 1940* (Oxford: Oxford University Press, 2010)

Schulting, Sabine, *Dirt in Victorian Literature and Culture: Writing Materiality* (London: Routledge, 2016)

'See-Saw, n. and Adj.', *Oxford English Dictionary* (Oxford: Oxford University Press, 2006)

Seifalian, Sophie, 'Gardens of Metro-Land', *Garden History*, 39.2 (2011), 218–38

Shepherd, John, 'Lansbury, George (1859–1940), Leader of the Labour Party', *Oxford Dictionary of National Biography* (Oxford: Oxford University Press, 2004)

Simpson, Michael, 'Adams, Thomas (1871–1940), Town and Country Planner', *Oxford Dictionary of National Biography* (Oxford: Oxford University Press, 2004)

Sini, R., 'The Social, Cultural, and Political Value of Play: Singapore's Postcolonial Playground System', *Journal of Urban History*, 48.3 (2022), 578–607

Snape, Robert, 'Juvenile Organizations Committees and the State Regulation of Youth Leisure in Britain, 1916–1939', *Journal of the History of Childhood and Youth*, 13.2 (2020), 247–67

Snape, Robert, 'The New Leisure, Voluntarism and Social Reconstruction in Inter-War Britain', *Contemporary British History*, 29 (2015), 51–83

Snape, Robert, and Helen Pussard, 'Theorisations of Leisure in Inter-War Britain', *Leisure Studies*, 32 (2013), 1–18

Spencer-Wood, Suzanne, and Renee Blackburn, 'The Creation of the American Playground Movement by Reform Women, 1885–1930', *International Journal of Historical Archaeology*, 21 (2017), 937–77

Springhall, John, 'Brabazon, Reginald, Twelfth Earl of Meath (1841–1929), Politician and Philanthropist', *Oxford Dictionary of National Biography* (Oxford: Oxford University Press, 2004)

Stark, James, *The Cult of Youth: Anti-Ageing in Modern Britain* (Cambridge: Cambridge University Press, 2020)

Steedman, Carolyn, *Childhood, Culture and Class in Britain: Margaret McMillan, 1860–1931* (New Brunswick: Rutgers University Press, 1990)

Steedman, Ian, 'Wicksteed, Philip Henry (1844–1927), Unitarian Minister and Economist', *Oxford Dictionary of National Biography* (Oxford: Oxford University Press, 2004)

Stewart, W.A.C., *Progressives and Radicals in English Education 1750–1970* (London: Macmillan, 1972)

Swenarton, Mark, 'Tudor Walters and Tudorbethan: Reassessing Britain's Inter-War Suburbs', *Planning Perspectives*, 17.3 (2002), 267–86

Swenarton, Mark, Tom Avermaete and Dirk van den Heuvel, eds., *Architecture and the Welfare State* (London: Routledge, 2015)

Tebbutt, Melanie, *Being Boys: Youth, Leisure and Identity in the Inter-War Years* (Manchester: Manchester University Press, 2012)

The British Academy, *Reframing Childhood Past and Present: Chronologies* (London: The British Academy, 2019)

Thomson, Mathew, *Lost Freedom: The Landscape of the Child and the British Post-War Settlement* (Oxford: Oxford University Press, 2013)

Thomson, Mathew, *Psychological Subjects: Identity, Culture, and Health in Twentieth-Century Britain* (Oxford: Oxford University Press, 2006)

Thorsheim, Peter, 'Green Space and Class in Imperial London', in *The Nature of Cities*, edited by Andrew C. Isenberg (Rochester, NY: University of Rochester Press, 2006), pp. 24–37

Thorsheim, Peter, 'The Corpse in the Garden: Burial, Health, and the Environment in Nineteenth-Century London', *Environmental History*, 16 (2011), 38–68

Tisdall, Laura, 'State of the Field: The Modern History of Childhood', *History*, 107.378 (2022), 949–64

Tomes, Jason, 'Amherst, Alicia Margaret, Lady Rockley (1865–1941), Garden Historian', *Oxford Dictionary of National Biography* (Oxford: Oxford University Press, 2004)

Tozer, Malcolm, 'A History of Eton Fives', *The International Journal of the History of Sport*, 30.2 (2013), 187–9

Voce, Adrian, *Policy for Play: Responding to Children's Forgotten Right* (Bristol: Bristol University Press, 2015)

Waites, Ian, '"One Big Playground for Kids": A Contextual Appraisal of Some 1970s Photographs of Children Hanging out on a Post-Second-World-War British Council Estate', *Childhood in the Past*, 11.2 (2018), 114–28

Walton, John K., *The British Seaside: Holidays and Resorts in the Twentieth Century* (Manchester: Manchester University Press, 2000)

Whalley, Robin, and Peter Worden, 'Forgotten Firm: A Short Chronological Account of Mitchell and Kenyon, Cinematographers', *Film History*, 10.1 (1998), 35–51

Whitfield, Matthew, 'Keay, Sir Lancelot Herman (1883–1974)', *Oxford Dictionary of National Biography* (Oxford: Oxford University Press, 2004)

Wildman, Charlotte, *Urban Redevelopment and Modernity in Liverpool and Manchester, 1918–1939* (London: Bloomsbury, 2016)

Willan, Jenny, ‘Revisiting Susan Isaacs – a Modern Educator for the Twenty-First Century’, *International Journal of Early Years Education*, 17 (2009), 151–65

Williams, Jean, *A Game for Rough Girls? A History of Women’s Football in Britain* (London: Routledge, 2003)

Winder, Jon, ‘Children’s Playgrounds: “Inadequacies and Mediocrities Inherited from the Past”?’, *Children’s Geographies*, 2023, 1–6 https://doi.org/10.1080/14733285.2023.2197577 [accessed 12 July 2023]

Withagen, Rob, and Simone R. Caljouw, ‘Aldo van Eyck’s Playgrounds: Aesthetics, Affordances, and Creativity’, *Frontiers in Psychology*, 8.1130 (2017) https://doi.org/10.3389/fpsyg.2017.01130 [accessed 9 Feb. 2024]

Wooldridge, Adrian, *Measuring the Mind: Education and Psychology in England, c.1860–c.1990* (Cambridge: Cambridge University Press, 1994)

Woudstra, Jan, ‘Danish Landscape Design in the Modern Era (1920–1970)’, *Garden History*, 23.2 (1995), 222–41

Woudstra, Jan, ‘Detailing and Materials of Outdoor Space: The Scandinavian Example’, in *Relating Architecture to Landscape*, edited by Jan Birksted (London: E & FN Spon, 1999), pp. 53–68

Wright, Valerie, ‘Making Their Own Fun: Children’s Play in High-Rise Estates in Glasgow in the 1960s and 1970s’, in *Children’s Experiences of Welfare in Modern Britain*, edited by Siân Pooley and Jonathan Taylor (London: University of London Press, 2021), pp. 221–46

Zweiniger-Bargielowska, Ina, ‘Building a British Superman: Physical Culture in Interwar Britain’, *Journal of Contemporary History*, 41 (2006), 595–610

Index